Winning Telco Customers Using Marketing Databases

For a complete listing of the *Artech House Telecommunications Library*,
turn to the back of this book.

Winning Telco Customers Using Marketing Databases

Rob Mattison

Artech House
Boston • London

Library of Congress Cataloging-in-Publication Data
Mattison, Rob.
    Winning telco customers using marketing databases/ Rob Mattison.
        p.   cm.—(Artech House telecommunications library)
    Includes bibliographical references and index.
    ISBN 1-58053-036-2 (alk. paper)
        1. Telecommunication — Marketing.   I. Title.   II. Series.
    HE7631.M376   1999
    384'.068'8—dc21   99-36699
                                                                                        CIP

British Library Cataloguing in Publication Data
Mattison, Robert M.
    Winning telco customers using marketing databases.—
    (Artech House telecommunications library)
    1. Telecommunication — Marketing
    I. Title
    384'.0688
    ISBN 1-58053-036-2

Cover design by Elaine Donnelly. Edited and illustrated by Brigitte Kilger Mattison.

© 1999 Artech House, Inc.
685 Canton Street
Norwood, MA 02062

International Standard Book Number: 1-58053-036-2
Library of Congress Catalog Card Number: 99-36699
10 9 8 7 6 5 4 3 2 1

I would like to dedicate this book to my good friend, Paulo Costa, whose support and insight helped make this book possible.

Contents

Preface

There is a battle going on today; make no mistake about that. As we enter the twenty-first century, this battle being waged will have a bigger impact on the personal, social, economic, and political world than any other in the history of humanity. This conflict, however, will not be fought with weapons of destruction and no one will be physically injured.

This battle is for dominance of the telecommunications environment of the future. Many skirmishes have already occurred in this ongoing conflict. Companies are fighting for supremacy in many areas. They fight to make their network infrastructures the newest, the most efficient, and the most robust. They struggle to improve their operational capabilities and their organizational excellence. And, they all realize that excellence in marketing and the pursuit and capture of the most and best customers is key to any strategy.

The marketing WAR

Unfortunately, most telcos are ill equipped to compete on the playing field of open competition. The industry was born and has grown in a relatively noncompetitive environment. It is only now learning the new realities that an open market telecommunications industry demands.

The major strategies in the telcos' adjustment are the concepts of WAR:

- *W* for wallet share: maximizing the revenue and profit delivered by each customer;

- *A* for acquisition: learning how to attract quickly and economically the best, most viable customers;

- *R* for retention: making sure that the customer, once attained, remains loyal to the firm.

This work, then, is a book about waging war in the telecommunications world and about how telcos are learning to make use of the marketing database as their key weapon in this war.

Rob Mattison

Acknowledgments

A book like this could never be completed without the support and hard work of many people.

First and foremost, I would like to thank Dr. Paulo Costa for his invaluable assistance in the preparation of this book. Dr. Costa's extensive background in statistics, analytics, marketing, and telecommunications was invaluable and without his hard work, his willingness to wrestle with complicated, confusing, and contradictory information, and his help in formulating simple, easy-to-understand concepts this book could not be as clear and useful as I believe it is.

I also want to thank my copyeditor, illustrator, and publication manager, Brigitte Kilger Mattison. Brigitte is the glue that holds the book together and her attention to detail and her sensitivity to the reader (forcing us to make the messages as simple, clear, and easy to understand as possible) is a crucial factor in its usefulness to you.

I also thank the SPSS Corporation, maker and distributor of a fine line of statistical analysis and data mining tools, and in particular Mr. Doug

Dow, director of strategic partnerships at SPSS, who supplied us with copies of software so that we were able to create the detailed explanations and screen shots included in this book.

Thanks are due my many coworkers and customers without whose kind interactions I would not be able to write as authoritatively. Special thanks go to Edward "Ted" Ross, Bill Gregory, Inder Koul, Michael Fuller, Don Cochran, Elsie Mustaller, Kevin Kulsavage, and a host of other people who contributed in some way to this work.

Introduction

Welcome to the world of telecommunications marketing

If you picked up this book, the odds are very good that you are an employee, a consultant, or a supplier to a company involved in the telecommunications industry. You are undoubtedly associated with or interested in the marketing activities of that organization. Most importantly, I am absolutely certain that, if the first two are true, you are also quite confused about the exact marketing process and how it is supposed to work in a telecommunications environment.

Why a book like this is essential

The marketing process in the majority of telecommunications firms today is hopelessly contradictory and seemingly counterproductive. In the past several years, the telecommunications industry in general and the majority of the companies in this industry have undergone a series of revolutions that have turned the world as we have known it on its ear. In just a few short years, this industry has been transformed from the stable, benevolent, government-sponsored and -supported providers of simple telephone service to consumers in a few concentrated markets to the biggest, most dynamic, most unstable, and frighteningly complex industry, responsible for delivering a myriad of voice, data, text, image, and other multimedia communications services to people from one end of the globe to the other.

Telecommunications is redefining the world as we know it

All over the world the telecommunications industry is in the process of redefining reality as we understand it. Think of the Internet, wireless telephones, satellite communication, pager technology, fax technology, cable television, ISDN, T1, ADSL, and so on. It seems that every day a new kind of telecommunications product is invented and yet another industry is formed.

Telecommunications is reaching all corners of the world. In China up to one million people became new subscribers to wireless telephone services in only one short month. On personal computers around the world, people are chatting and watching each other with Internet video technology. Powerful new telecommunications networks enable the business world to redefine how things get done.

Telecommunications is now considered to be a human right

Telecommunications services have become so pervasive and so crucial to the survival of an economy and a country that the United Nations has proposed that the right to telecommunications access be recognized as a basic human right to be afforded to all peoples. In fact the ITU, the International Telecommunications Union (the organization that provides a telecommunications voice and positioning for the United Nations), has declared this to be the theme for their 1999 telecommunications conference in Switzerland.

The role of marketing in telecommunications

This is all very exciting, but how does marketing fit into it? Fortunately, marketing of telecommunications plays a critical role in the organized, efficient, and rapid deployment of these new (and old) technological capabilities to companies and individuals everywhere. Although telecommunications will continue to be an "engineering-driven" industry for some time to come, the role of marketing is increasing in importance.

Why marketing is so crucial

Unfortunately, many people have a somewhat jaundiced and unflattering view of the marketing process and the role of the marketer. They see marketing as trivial, frivolous, and unnecessary, a drain of good resources, and, in some cases, an expression of vanity. In their eyes, marketing is a necessary evil that gets in the way of the true work of telecommunications firms, which is to build better networks and deliver more messages. More enlightened individuals, however, realize that the success of any of these new approaches will not depend solely on how well they are engineered, but how well they are understood and accepted by the public and business communities.

Marketing is crucial because without it the whole process of technological development and deployment would be stalled (as it was in the past when telecommunications firms were protected and not forced to compete).

The competitive imperative

When the governments around the world decided that the well-protected telecommunications monopolies of the past needed to be spurred into new efficiencies through open competition, they basically mandated that marketing (packaging, presenting, and promoting of telecommunications products and services) would need to become a key component of any telecommunications company's survival strategy. Indeed, without marketing, there would be no competition and hence no reason to deregulate.

Like it or not, you *must* market

Other industries were allowed to find their own comfort level in the use of marketing. The telecommunications industry, however, is literally being *forced* to learn how to market, and learn it quickly, whether the industry wants to or not.

To succeed, you must market well

As soon as a government opens up a market and allows the forces of competition to come into play, two teams immediately take to the playing field: *incumbents*, the companies that were there from the beginning (and the "owners" of 100% of the existing market), and *new entrants*, companies with 0% of the existing market. In other words, when a market is deregulated, the government basically pre-declares a group of winners (the new entrants) and losers (the incumbents). Imagine!

From that point forward it is a race to figure out how to market, who your customers are, which of them are good customers and which are not, and how to keep the good ones. Suddenly, the telecommunications firm must grow aggressive, market-sensitive appendages that it never had before.

The role of the marketing process

It should come as no surprise that telecommunications firms are a bit chaotic and schizophrenic when it comes to figuring out the marketing aspects. After all, while industries such as finance, retail, and manufacturing have been competing in an open market for a long time, the poor telecommunications marketer has only a decade or so of true experience from which to draw, and that from only a handful of countries (the United States, United Kingdom, and Chile) who pioneered the open market concepts only recently. Consequently, most telcos are only in the early stages of developing a truly robust model of what marketing is and how it should work in the telecommunications context.

Developing a model of the marketing process

This book proposes a model for what the marketing process is and how it can (should/might) work in a mature telecommunications marketing area. This model is a unique combination, borrowing marketing practices and frameworks developed in more mature industries, customized to address the specific conditions that most telecommunications firms are facing.

The role of the marketing database in telecommunications

Along with the marketing process, we will describe the role that marketing databases can and will play in the telecommunications marketing environment. From the early competitive pioneers of telecommunications marketing we have learned that the marketing database is a key component of any successful marketing formula. We will, therefore, spend considerable time talking about how the marketing database is designed and built, used, staffed, and fit into the marketing process.

Who should read this book

Who then should be reading this book? Actually, we have targeted its contents toward many people.

Marketing executives and practitioners

The most obvious group of people to whom this book is directed are those actually involved in managing and executing the marketing process in telecommunications firms. These are the people in the trenches who face the organizational and operational challenges we talk about. For this group we hope to provide a framework for understanding the process they participate in a little bit better and a set of guidelines for making the marketing process more efficient, easier to understand, and a more pleasant experience for everyone.

Computer systems developers and supporters

The second and equally important group of readers for this book consists of the people involved in the construction and support of marketing computer systems in the telecommunications firm. They are the individuals who often feel that they are the tail that is furiously wagged by the dog (namely, the ever-so-demanding and constantly changing marketing executives).

We hope that this book, and the framework it presents, will provide this group of professionals with the detailed information needed to help them understand the marketer's needs in both the short and long term, and the guidance necessary to construct powerful, flexible, and cost-effective marketing database systems.

Marketing modelers

The third constituent in the world of marketing execution and marketing databases is the analysts. Marketing executives are in charge of setting the strategy and choosing direction. The computer systems professionals take care of finding the data and building the systems. It is the analysts, however, that make the modern telecommunications marketing organization an effective agency. These analysts are the modelers, segmenters, statisticians, and other numbers crunchers.

This book establishes a context for these analysts to better explain what they do to the other members of the marketing team. It also provides a vocabulary and a structure that allows the three groups to communicate and coordinate their activities effectively.

Advertising agencies and direct marketing firms

Although the marketers, computer technicians, and analysts are the primary audience for this book, several other groups will find it helpful as well. It can serve as a road map for the myriad advertising agencies, direct marketing firms, and other marketing services outsourcers interested in assisting telecommunications accounts. The book can provide you with an in-depth understanding of what telco people are dealing with.

Telecommunications executives

If you are the "mahogany row" executive of a telecommunications firm, this book can provide a useful template for understanding the current role of the marketing organization in your firm and what it might be in the future.

AH Artech House Publishers BOSTON · LONDON

To receive information on new and forthcoming titles from Artech House, please fill out the other side of this card and mail or fax it to one of the locations below:

For Europe, Scandinavia, the Middle East, Africa:

Artech House Books
Portland House, Stag Place
London, SW1E 5XA U.K.
+44 (0)171 973-8077
FAX: +44 (0)171 630-0166
artech-uk@artech-house.com

All other regions:

Artech House Publishers
685 Canton Street
Norwood, MA 02062 U.S.A.
(781) 769-9750
1-800-225-9977 (continental U.S. only)
FAX: (781) 769-6334
artech@artech-house.com

Find us on the World Wide Web at:

www.artech-house.com

AH Artech House Publishers BOSTON · LONDON

As a buyer of Artech House technical and professional books, you are invited to receive the latest information on our new and forthcoming titles. To be included on our mailing list, fill in the address details below, noting your areas of interest and return to one of the locations listed on the back of this card. We look forward to your reply.

Name: _____

Position: _____

Company: _____

Address: _____

_____

Tel: _____

Fax: _____

E-mail: _____

Please indicate your areas of interest:

- ☐ Radar
- ☐ Microwave
- ☐ Optoelectronics
- ☐ Remote Sensing
- ☐ Computer Science
- ☐ Computer Systems & Architecture

- ☐ Software Engineering
- ☐ Signal Processing
- ☐ Solid-State Technology & Devices
- ☐ Antennas
- ☐ Telecommunications Engineering

- ☐ Telecommunications Management
- ☐ Technology Management & Professional Development
- ☐ Other

Thank you!

Hardware, software, and consulting sales organizations

Finally, the information in this book can, of course, be used to guide the manufacturer and provider of marketing support hardware, software, and consulting services.

How this book is organized

As stated, marketing in the telecommunications industry is chaotic and confusing. In this book, we explain, organize, and demystify much of that chaos.

The book is organized into five major sections, and each is further broken down into individual chapters of interest. The following provides you with a brief synopsis of the sections and chapters.

Part 1: An introduction to telecommunications marketing

In the first part, we introduce the reader to the concept of marketing and its unique manifestation and role in the telecommunications environment. We also establish our *telecommunications marketing framework*, the road map we utilize to explain various aspects of the marketing process.

Chapter 1: We don't need to market—we are the phone company!

In the first chapter, we consider the history of telecommunications and the role marketing plays. Included is a look at the current worldwide trend toward deregulation versus monopoly-based telecommunications approaches. We consider the strengths and weaknesses, and the reasons why governments choose one model over another and what telcos can subsequently do about it.

Chapter 2: An introduction to telecommunications marketing

Here, we introduce the concept of the *marketing process*, a well-defined, easily understood framework that explains how the different pieces of the marketing puzzle in a telecommunications firm fit together.

We establish the critical role that marketing plays in the telco, namely, to stay aware of the many different forces (competition, market, regulation, innovation, and so on) that affect the direction the firm should take, and to provide leadership in setting that direction and in creating the specific messages sent to the public via various campaigns.

We also introduce the concept of the marketing campaign itself, and explain the five dimensions that define it, the five M's: merchandise, market, message, media, and margin.

Chapter 3: The core marketing process

Having established a basic background for understanding the marketing process in Chapter 2, we turn in Chapter 3 to the actual definition of the process itself. We define the four steps or phases of the marketing process (prioritization and goal setting, modeling, campaign development, and campaign execution) and we describe the organizational elements that support it (the people, procedures, techniques, technologies, and marketing databases).

Part 2: Understanding marketing campaigns

Armed with an understanding of the basic components of the process, and the process itself, we turn in Part 2 to a more detailed discussion of the five M's of campaigns.

This part is concerned with the many different considerations of putting together a marketing campaign. We see the effects that the use of different media, the selection of different target markets, the preparation of different messages, and the sales of different kinds of products and services have on the ability of the marketer to perform effectively and on the marketing database to support them in that effort.

Chapter 4: Telecommunications strategy and campaigns

In this chapter, we provide the reader with a detailed exploration of the strategic perspective a telecommunications firm has and the role that marketing and the execution of specific campaigns play in that strategy. We also look at some of the most successful campaigns that U.S. telecommunications firms have run and diagnose them in terms of the strategic objective.

Chapter 5: Media, messages, and outsourcing

To the outsider, the process of marketing might seem to be a single, homogenous environment. In reality, there are three different basic media the telecommunications marketer can use: mass, direct, and promotional media. These media and the messages to be delivered are very different in terms of how they are used, their cost, and how they are managed. In this chapter, we consider the similarities and differences and explore how they can impact the builder of a marketing database system. We introduce the *Costa model,* which shows the different levels of awareness and acceptance that lead to better customer relationships and ultimately to more sales.

Chapter 6: Direct marketing

Here we review the fundamentals of the direct marketing process. We look at its history, its progress, and its current position of power in telcos. Indeed, most telcos are finding that direct marketing, much more than mass marketing, is the key to success in long-term and strategic planning.

We also describe the fundamental steps in the direct marketing process and the critical role that the marketing database plays in making that process work.

Chapter 7: Advertising and promotion

Chapter 7 focuses on the other media that the telecommunications marketers have at their disposal, mass media (advertising) and promotion. In this chapter, we see how very different the entire process of choosing target markets, preparing messages, and testing results really is, and we develop an appreciation for the challenges one faces when trying to measure the effectiveness of these kinds of activities.

Chapter 8: Acquisition, retention, and wallet-share campaigns

Finally, in Chapter 8, we zero in on the three foundational objectives of all campaign activities: *acquisition,* gaining new customers; *retention,* maintaining existing customer relationships; and *wallet share,* improving the revenue/profit that the company gets from existing customers. We examine the fundamentally different nature of these three kinds of campaigns and look at some of the approaches that telcos use to prioritize, model, develop, and execute campaigns.

Part 3: Analytics

Now we are ready for a detailed discussion about analytics, the statistical analysis process that goes on behind the scenes of every marketing activity. In this section, we examine analytics from several perspectives, including simple reporting analysis, and the more advanced forms of segmentation analysis, scoring, and the development of functions.

If you want to understand the modeling process, and if you want to have a context for how detailed statistical and data mining analysis fits into the overall planning and execution of campaigns, then this part has the information you need.

Chapter 9: Product, customer, and competitive analysis

In this chapter, we examine the challenges and nature of reporting and analysis problems in the telecommunications area. We look at the kinds of product, customer, and competitive information that companies typically work with, and we examine some of the most common reporting structures used that make that information available to marketers.

Chapter 10: Simple segmentation

Chapter 10 introduces the concept of segmentation. We explain what it is, how it is used, and why it is so important. We also clarify the three major types of segmentation models, namely, the simple/structural, complex/characteristic, and behavioral models, and explore the concept of the structural segment in detail.

Chapter 11: Complex segmentation

In this second chapter about segmentation, we look at the most popular and most easily recognized form, the characteristics-based segmentation study. Characteristics-based segmentation models attempt to describe customers and divide them into groups according to descriptive characteristics. Demographics, psychographics, and most of the other popular segmentation designations are of this type. In this chapter, we describe how these kinds of models are conceptualized and review the different tools that analysts use to develop them.

Chapter 12: Behavioral segmentation

Here, we review the most powerful and least understood area of segmentation, behavioral segmentation. We learn that characteristics-based segmentation is useless without some kind of base of behavioral information from which to draw. We also examine the different sources of behavioral data for the telecommunications marketer and the various ways in which behavioral models can be generated.

Chapter 13: Scores and functions

In this final chapter on analytics, we look at two other forms of statistical analytics used with great success by the marketer. We explore the creation of scores that are used to rank or rate customers according to a complex set of criteria and functions, which attempt to summarize complex relationships between different characteristics of the customer's behavior and valuation into a simple, easy-to-understand numeric identifier.

Part 4: The marketing process

Having come to understand campaigns themselves and the analytics that support them, we reexamine the marketing process to see how truly rich and complex it actually is. Armed with this detailed perspective, we are then able to look at the computer systems that can be used to support them.

Chapter 14: Prioritization and goal setting

This chapter covers the process of sponsorship, goal setting, and the creation of marketing teams. A good job of managing this part of the marketing process greatly reduces the inherent chaos of the rest.

Chapter 15: The modeling process

In this chapter, we summarize what we have learned about the types of models that a marketer can develop (Part 3) and the types of campaigns that can be executed (Part 2). We see how the modeler can pull these pieces together into a logically and mathematically sound proof and plan of action.

Chapter 16: Campaign development and execution
Finally, we review the effects on the marketing organization and on the marketing database of trying to merge mass, direct, and promotional campaigns into one organizational, procedural, and executional framework.

Part 5: The marketing database
This is by far the most technical section of the book. Building on what we have learned in the earlier sections, we create a comprehensive view of how a marketing database system can be built to best manage the marketing process.

Chapter 17: Marketing database functionality
In this chapter, we analyze the many functional requirements that the different components, processes, and media place on the marketing database designer and define the various categories of functions it must support. We define the three major subsystems of any marketing database: the *query and reporting*, *analytics*, and *process management* subsystems, along with the major sources of data used to support them.

Chapter 18: List management
In Chapter 18, we take an in-depth look at the number one, most often automated of the telecommunications marketing processes, the direct marketer's *list management* process. We see how list management is done, the critical business role it fulfills, and how such a system can be built or selected.

Chapter 19: Campaign management
In this chapter, we examine the area of the marketing process that most people would like to see computer managed (although few actually do), the *campaign management* process. Here we examine the requirements and some alternative approaches to making campaign management a truly

effective process across all media, time frames, segments, and other challenges.

Chapter 20: Marketing database: architecture

Finally, we look at the marketing database hardware and software itself. We talk about how it can be built, how the different parts can be constructed, and how the supporting databases can be defined and delivered.

Part 1

An Introduction to Telecommunications Marketing

1

We Don't Need to Market—We Are the Phone Company!

It was a cold winter's day. We were sitting in the executive meeting room of the nation's only telephone services provider. Outside the window, we saw picturesque snow-covered hillsides and a beautiful centuries-old cathedral. On the table sat our proposal for the development of a new marketing database system. The audience listened politely as we presented each of the points covered in our plan. Several members seemed more interested in the scenery than in our presentation. Finally, we came to the end.

"How much will the system cost?" asked the chief financial officer.

"Well that all depends," I replied, practicing my best noncommittal consulting speak.

"Never mind the conditions," she barked! "Just get to the bottom line—*how much?!*"

It was obvious that I would have to deliver some numbers, so I gave them to her.

The explosion of exhaled air almost created a wind in the room. The looks on the faces of most of the participants grew decidedly colder than the weather outside, and I thought about how much more fun and less dangerous skiing down the slopes would be right now.

"Why that's ridiculous!" I heard someone say. "We don't need to spend that kind of money on marketing. After all, we are not selling shoes or concert tickets, *we are the phone company!*"

We are the phone company!

Since the 1980s, scenes like this have been playing themselves out in the ranks of telecommunications companies around the world. It seems that a very dirty trick has been played on the executives of most telecommunications companies. The rules by which telephone companies have been playing for more than 100 years have been changed. A little history will help to understand the significance of such a statement and what it has to do with marketing strategies for telecommunications firms today.

U.S. telephony foundations

On March 7, 1876, U.S. Patent No. 174,465 for a fully functioning telephone was issued to Alexander Graham Bell. This invention, the first of its kind to display the characteristics of "harmonic telegraphy" (sending sounds over telegraph wires), became the foundation of an industry that has served to continuously improve the standard of living and speed the rate of change in our world ever since. Ironically, nobody showed much interest in this invention. For more than a year, Bell and his partner, Thomas A. Watson, survived by putting on telephone side shows where people paid to hear music sent over the wires from one room to another. Eventually, of course, interest grew and the Bell Telephone Company was formed.

Early competition in the United States

Although other countries may have a slightly different history, the story of the progression of the telephone business in the United States is certainly illustrative. From 1877 when the Bell Telephone Company was founded until 1910, the U.S. telecommunications industry functioned in a dual mode of operation. The Bell Telephone Company thrived as the largest driving force behind the industry. It maintained control over the lion's share of the local telephone lines and the manufacture of telephone equipment. (In 1882 the Bell Telephone Company purchased Western Electric Company, the largest manufacturer of telephone equipment at the time.) It also maintained exclusive control over long-distance services. But the corporation also had its struggles.

Telephone versus telegraph

The Bell Telephone Company was far from the only game in town. The biggest initial challenge to control of the telephone industry in the United States came from Western Union. Western Union was the already established owner of the U.S. telegraph business with an extensive network of national scope in place. With acquisitions of patents on telephones created by Thomas Edison and other inventors, Western Union created a telephone business that competed directly with Bell. This fierce competition lasted for 7 years until the managers of Western Union realized that they were destined to lose a patent infringement case and decided to settle out of court in 1879. The Western Union settlement gave Bell control over the largest single telephone company of its time. In 1885 the American Telephone and Telegraph Company (AT&T) was formed to provide nationwide telephone service.

Competition abounds

Although Bell was in control of the largest phone company in the United States, he in no way had control over the entire industry. By the time Bell's patents expired in the mid-1890s there were more than 125 "independent" telephone companies (that is, small telephone service providers not associated with the Bell companies) and several large suppliers of telephone equipment. AT&T's competitive response to these companies was simple and straightforward:

- To bring suit against the companies for patent infringement;
- To offer much lower rates than the competition;

- To buy out competitors;

- To refuse to connect these small independent companies to the customers serviced by the Bell companies.

Of course, this kind of policy, especially one to refuse connection services, was bound to get government attention and in 1910, the U.S. government cited AT&T for violation of antitrust laws. To prevent Justice Department action, AT&T issued a letter called the "Kingsbury Commitment," which promised that AT&T would stop buying up competitors, and would provide access to its customers for all competitive telecommunications companies. It wasn't until 1921, when Congress passed the Graham Act that exempted telephony from antitrust laws, that the AT&T monopoly over telecommunications was legitimized.

The days of monopoly in the United States

In 1921, the U.S. government passed a law that basically sanctioned Bell to run a monopoly for telecommunications services in the country. During that time, AT&T had grown into one of the largest corporations in the world and was by far the largest carrier of telephone services. For more than 50 years, AT&T enjoyed a relatively free hand to run the telecommunications industry in the United States as it saw fit. There were still a respectable number of competitors, but these companies were for the most part small, disorganized, and in control of rural and remote markets that AT&T did not want to manage anyway.

Telephone service in the United States continued to grow in effectiveness and efficiency and for a while it seemed that the perfect mode of operation for telecommunications had been realized.

The fall of the AT&T dynasty

Although AT&T continued to increase its market share and consolidate its position, the basic forces of monopoly economics continued to play out. The monopoly that AT&T held over this lucrative industry did not please everyone. The 1960s and 1970s saw an increasing sense of anxiety on the part of consumers and government officials as the telephone continued to grow in importance. Jokes about the phone company being more powerful than the government and about secret telephone conspiracies became commonplace. In fact, a comic movie called *The President's Analyst,* starring James Coburn, told the story of a world where the telephone company turned out to be the single most powerful, clandestine government on the earth.

Along with this growing public sense of unease over the monopoly came hundreds of technological innovations. Thousands of new opportunities for the expansion of voice and data transmission in economical ways were being invented and entrepreneurs were clamoring to exploit these new markets. Of course, that whole industry was owned by AT&T and it was not about to give up any of that market without a fight.

As a result, several lawsuits, including the Carterfone suit of 1968 and the MCI case of 1969, saw AT&T accused of stifling competition and of preventing the full exploitation of the economic possibilities offered by new alternative telephonic solutions. (Carter Electronics used telephone lines to connect two-way private radio messages, and MCI sent voice messages via microwave signals.) AT&T and the policy of monopoly lost both cases and the beginning of the end of the telephone monopoly in the United States had begun.

In 1974, the government filed a new antitrust suit against AT&T calling for the local, long-distance, and manufacturing facilities to be separated. The suit was terminated in 1982 with a consent decree issued by Judge Harold Greene which ordered this divestiture to occur.

Post-divestiture and the current state of U.S. telephony
The 1982 Greene decision heralded the end of monopolized telephony in the United States. Since that time the country has seen an explosion in the number of service providers, the kinds of services available, and the different rate plans and payment schedules available. In less than 20 years the U.S. telecommunications marketplace went from having one dominant provider of all local and long-distance services to dozens of long-distance, local, cellular, personal communication system (PCS), Internet, and pager services. At the same time, the amount of telephonic traffic skyrocketed far beyond the levels originally anticipated as corporate data, video, audio, Internet, and other kinds of signals filled the wirelines in ways never before imagined.

Deregulation around the world

Although the U.S. telephone history is interesting, it does not represent the full story. Similar and equally tumultuous accounts can be told of the advancement of telecommunications around the world. In most countries telephone services started around the same time as in the United States. Of course, different countries had different standards and laws in place, but for the most part the telephone business followed the trails blazed by the developers of telegraph services. In some countries, the telephone company was simply another branch of the government. (This is still true in a few cases.) In other countries, open competition was allowed to reign. In still others, protected monopolies (private companies with special government-granted rights) were established.

By the mid-1900s however, every country on the planet was being serviced by a telephone company of one kind or another. By and large, these companies enjoyed a special relationship with the government that allowed them to function without much fear of competition.

Current rate of deregulation in the worldwide telecommunications industry

Today, however, the global telco marketplace is changing at a rapid pace. The same pressures that forced the U.S. government to reconsider its monopoly decision and to deregulate telecommunications are making themselves felt in nations around the world. More and more governments are deciding that the survival of their economies and the needs of their people are better served by a more competitive marketplace than they have allowed to exist until now. The facts speak for themselves.

Deutsche Telekom Ag (DT), Germany

Deutsche Telekom Ag, the German telephone company, completed its first initial public offering in November 1996 when it sold 26% of its shares (another 24% is set to be sold in 1999). Germany liberalized all nonvoice telephony in 1992 and the last monopoly has been undergoing transformation. As of the end of 1999, there were more than 24 recognized and declared competitors operating or set to open business within the German borders.

Nippon Telephone and Telegraph (NTT), Japan

Historically, the Japanese marketplace has been very heavily regulated. In 1985, the Ministry of Posts and Telecommunications formulated a plan to partially privatize the telephone industry in Japan and to foster competition. The market was traditionally dominated by NTT with more than 72% of the local loop market and 99% of the domestic market. This plan calls for disengagement of NTT into one long-distance and two regional domestic phone companies (a plan similar to the Greene plan in the United States).

Telecomunicacoes Brasileiras (TELEBRAS), Brazil

The Brazilian government struggled with telco privatization for several years. In 1998, the much heralded "auction" of telecommunications sectors in the country saw the infusion of large amounts of capital and much needed expertise as worldwide telco concerns were allowed access to the largest and most lucrative of the Latin American telecommunications markets.

British Telecom (BT), England

Following closely behind the United States, privatization in the British market began in 1984, and voice-based telephony was fully deregulated by 1996 with more than 150 carriers now licensed as service providers. Despite this environment, however, BT still maintains control of more than 80% of the market.

The telecommunications marketplace today

In country after country telecommunications has followed a similar path and is continuing down very similar and often convergent paths as we prepare for the twenty-first century. Telco companies are getting larger and are beginning to cross national boundaries in ways never before imagined. The recent efforts to privatize and modernize telephone companies around the world have seen the merger, acquisition, and partnering of telcos from one end of the world to the other. Companies such as AT&T, BT, Telfonica (the Spanish phone company), and the national phone companies just about everywhere else around the world are joining to form newer, larger, more efficient, and more profitable telecommunications companies than ever before.

Understanding competition in telecommunications

So, the telecommunications industry is undergoing a major shakeup. Competition is being foisted on telcos like never before. What does all of this mean? What impact does this competitive atmosphere have on telcos, and what are the appropriate responses?

Why is competition a bad thing for telecommunications?

Before we celebrate what a great deal deregulation is for the consumers of telco services, let's consider some of the disadvantages. There are actually several good reasons for countries to try to keep the telco industry as a government-regulated monopoly and, conversely, there are several downside issues we need to consider.

Underwriting the cost of infrastructure

One of the major motives for a country to protect its telecommunications industry is the same motive used when governments accept responsibility for building roads and establishing schools. It is very expensive to set up a national telephone system and those costs are generally far greater than the benefits a company can reasonably expect in the short or medium term. Before they could open for business, they would have to purchase the right of way to provide wire access to every single home, business, and government building in the entire country. Then they would have to install poles and lines to carry the traffic. After that, switches, operators, and other kinds of infrastructure and support services would have to be installed.

Let's use Australia as an example to provide us with a picture of the challenges that governments face. Wiring up a metropolitan area like Sydney would be extremely profitable and finding a company interested in servicing that market would be no trouble at all. To provide similar services for the people in, say, Alice Springs, a city located in Australia's outback, thousands of miles from a major metropolitan area, with a population of just 27,000, is an entirely different matter. Obviously, far fewer companies would be interested in providing service for that market.

A telephone system that only connects some of the people some of the time, however, is not an acceptable system and presents governments with a problem. The solution is clear in most cases—allow a company to have access to the profitable, dense markets (such as Sydney) and require that they use some of the profits to finance the remote locations (such as Alice Springs).

Development of a monopolistic environment allows very large numbers of people to enjoy a high level of service at a reasonable cost no matter where they are located. Granted, the telephone users in the major cities could get the same service for less, but the difference in those costs is actually minimal.

Connectivity issues

Underwriting infrastructure development costs, in and of itself, is a compelling reason for countries to stick with the monopoly model for telco, but there is another, equally valid reason: the connectivity issues involved. What was Bell Telephone's number one tool in curtailing its competitors' activities? What is the best way to make sure that another telephone company cannot benefit from your network? It's easy, just don't allow their customers access to your customers.

If competition is allowed to flourish in the telecommunications market, each telco will ultimately demand that all customers hook up via their network only. The result in openly competitive markets is that people need to get phone services from both or several companies just to get connected to everyone. That was certainly the case in the United States in the early 1900s and it was equally true in many other modern economies. For example, only a few years ago Mexican businesses needed to subscribe to two different phone companies to be able to send and receive calls from consumers throughout the country.

Standardization issues

Finally, there is the problem that countries experience when competing telcos install different kinds of telecommunications infrastructure equipment. Companies that install incompatible switches and wiring are guaranteeing that the cost of telecommunications services will be prohibitive in their marketplace.

So why have competition now?

If there are so many good reasons for countries to have a monopolistic form of telecommunications service, why then are so many of them converting to a much more aggressively competitive model? There are several reasons:

1. *Infrastructure in place.* First and foremost is the fact that a large amount of infrastructure is already in place and there is no reason for consumers to continue to underwrite the effort.

2. *Non-infrastructure-based products.* The second reason is that more and more telecommunications products are coming on the scene which require much less of an infrastructure investment than the original phone lines. Wireless and pager services have some hard network costs but they are not anywhere near as high as that of landlines. New, nontraditional offerings such as cable-based services allow phone services to be delivered over nontelephone lines.

3. *Standardization can be maintained.* The industry is so large and extensive that it would be very difficult to bust the standards as they exist today.

4. *Connectivity rights.* Nearly every legal body in the world has now established that open and full access to all subscribers to a service is part of the requirement that any telephone service provider must meet.

Telecommunications—multiple lines of business

The fact that a telephone company can be in many different lines of business today also makes it extremely difficult to maintain monopolistic practices. Telephone companies have gone from the simple providers of home-to-home voice service to multidimensional suppliers of a wide range of products and services including:

- Local voice telephone (the original sole product of the telephone company);

- Data transmission (both for businesses and homeowners);

- Paging;

- Wireless phone (via cellular, PCS, and satellite);

- Long distance;

- Internet hosting;

- ISDN and other "conditioned line" services.

The explosion of different products that a telecommunications company can offer has created yet another important aspect of the revolution telcos are going through.

The role of marketing in the telecommunications firm

In this highly competitive market it would be absurd to think that a telecommunications firm does not need to market! At the same time, it is clear that the role of marketing in the telecommunications firm is very different than in retail, manufacturing, or banking. Telecommunications is a unique industry with a unique history and characteristics that dictate what marketing can and cannot accomplish.

The goal of every telecommunications firm

To know how to position marketing in the framework of a priority setting it is important that we first establish the telecommunications firm's goals. This may appear to be a silly thing to think about. Isn't it obvious? Actually, no, not at all. The answer depends on the country where the firm is located and on the regulatory environment in which it is operating. For a telephone company that is still working in a monopoly or in a very protective environment the objectives of the firm will be set by the governmental agencies that regulate it. Usually, the mission of a state-sponsored telco is to provide a high level of service to all citizens at the lowest cost possible. If the telco exists in a highly competitive market, however, then the goal will be very different. In that case, the objective of the firm will be *to make a profit*.

Let's consider the differences in how one would run a telco in each of these situations and the different roles that marketing would play.

Goals and challenges for the noncompetitive telco

For the government-sponsored telco, the biggest challenges will be centered in the customer services area. This organization will be constantly vulnerable to the claims of various groups of people that the level of service they are getting is not up to the standards the government expects. Customer service and political maneuvering will be the principal means of dealing with customer satisfaction issues.

Aside from the vulnerability to political pressure is the issue of how to deal with the less desirable customers, those that have payment problems, for example, or are very expensive to provide service for, such as the remote customers. The government-protected business does not get to pick and choose customers the way another business can.

With such a business model, the last thing the company needs is any kind of marketing activity to reduce profits and increase demand. Telcos in this situation have all the customers they need, lucrative or not. These companies make profits by minimizing the exposure to undesirable customers and maximizing the efficiency of their operations. In a typical state-sponsored telco, engineering and operations are the disciplines that define how well the company does.

Goals and challenges for the competitive telco

If the telco is in a highly competitive marketplace, however, the scenario changes drastically. In this environment, a company has much more leeway in deciding which customers to deal with. Credit policies can be tightened and the company can get exceedingly selective about which markets and which customers to pursue. Under this business model, the situation is much clearer. There is no worry about political pressure, and profit, not political contingency, defines success.

Of course, if any telco is to survive, it still needs to have a good network and reliable customer service, but the priorities are shifted. In the competitive marketplace, the marketing discipline suddenly becomes a critical part of the profitability equation. Since this company can choose the market segments it deals with, the first thing it needs to know is which customers are the most desirable. Only marketing can answer that question. Next, the company needs to decide how to reach those people and how to make themselves an attractive alternative to the other telcos that are suddenly available. Again, the marketing discipline comes into play.

Perceptions of marketing in telecommunications

Unfortunately, this shift from the noncompetitive to the competitive marketplace is not an easy one for most telecommunications companies. First, few governments have been willing to give up complete oversight of and control over their national telecommunications markets. This means that government sponsorship and intervention are simultaneously mixed with open marketplace rhetoric and competitive pressures. The typical telco executive is forced to deal in a fuzzy half-world where both sets of rules apply at the same time.

Second, the cultural heritage of the vast majority of telecommunications firms, and especially the ranks of upper management, are inundated with beliefs, predispositions, and prejudices based on the old way of running a telecommunications business. This means that new ideas about the role of marketing and how things should be done in the firm often meet with stiff resistance from the people in authority positions. Indeed, the very structure of the telecommunications firm, the lack of integration between marketing and a marketing-based way of thinking, and the way other departments make decisions, all serve to undermine marketing efforts.

Why is the integration of marketing the key to success?

The incredible lack of integration between the marketing function and the other functions of the business is one of the most frustrating issues for the telco marketing practitioners to deal with. Remember that in most industries (especially retail, transportation, manufacturing, and services) marketing has always played a critical role in the success of the company. Talk to anyone in the retail, manufacturing, or services industry and they can describe to you in vivid detail exactly how marketing fits into the business and how it helps coordinate the activities of all other areas.

Can you imagine the manufacturer of a breakfast cereal successfully selling a product without the coordinated efforts of sales channels, distribution, market research, manufacturing, advertising, and a dozen other organizations. The marketing department coordinates all of those activities. Can you imagine an auto manufacturer who would even dream of designing and marketing a new car without the benefit of millions of dollars worth of market research, customer surveys, and other market readiness work?

In each case, the decisions made by the company and the actions it supports are not only accompanied by marketing activity, but are actually integrated and coordinated by marketing. Most importantly, are you aware of *any* telecommunications company that is capable of mustering even one-tenth of that much marketing activity around any of its marketing efforts? Probably not.

The big secret about marketing

The secret about marketing that most telecommunications firms have not yet realized is that the key to successful marketing is not how well the marketing itself is done. In fact, a very poor marketing plan can yield some very good results. No, the key to success is in how well the marketing function is used to coordinate the activities of all the different areas of the business around the objectives of meeting customer needs and creating the maximum amount of profit for the firm. Marketing, in and of itself, is a small part of the business's processes, but an incredibly important one.

In the next chapter we continue this discussion by taking a much closer look at what marketing really is and how it fits into the overall running of the telecommunications firm.

2

An Introduction to Telecommunications Marketing

A good beginning makes for a good ending.
English Proverb

If you don't understand the basics of the game,
you'll never be a winner.
Vince Lombardi

What is marketing anyway?

As we have already discussed, one of the biggest challenges for telecom-munications companies that want to get involved in the business of marketing is their very short history with marketing functions. They have very little to go on while they learn how the marketing functions will fit into their business process. The danger here is to look at other industries' marketing practices and assume that they are valid for the telco. Invariably, those practices either do not make sense or do not apply. The tendency then may be to totally disregard the discipline as being irrelevant and frivolous. That would mean, of course, that the useful, productive results that marketing can provide are dismissed as well.

Organizational challenges in the telecommunications industry

Marketing has had a long and impressive history in making companies successful in other industries. For most telecommunications firms, however, it is still a new and untested discipline. This causes some special problems as the sector begins to mature.

Marketing departments versus marketing functions

All telecommunications companies perform some kind of marketing functions, but they don't all organize them the same way. Some companies create large, formally defined marketing departments and require that all operational units and executives channel all marketing activities through those groups. Other firms establish the marketing department in more of a consulting role. The marketing staff provides support for the operational and executive managers in the design and execution of campaigns, but leaves the responsibility for budgeting and managing marketing activities to the individual business managers.

In many organizations, marketing develops a hybrid identity somewhere between these two extremes. This means, of course, that the acknowledged marketing department employees are not the only people involved in the support of marketing activities.

Some recognizable marketing activities

In summary, the marketing discipline (regardless of which individual or organization does the work), as it applies to most telecommunications firms, involves any functions that are concerned with:

- Discovering and developing new product offerings and packages;
- Developing an understanding of the profitability and marketability of existing product lines;

- Managing the relationship between customer and company, including strategic and coordination efforts in the areas of customer service, sales, advertising, and credit and billing in pursuit of marketing objectives;

- Developing and executing campaigns designed to accomplish specific measurable objectives in terms of market share, product utilization, profitability, or market penetration;

- Developing an understanding of the current competitive position of the firm and the setting of future trends, directions, and goals on the basis of that analysis.

Given this rather broad-reaching definition, let's look at how different telecommunications firms have made use of marketing to help them in each of these areas.

The need for a framework

The study of any subject as complex and far reaching as marketing can be confusing and misleading at first. Indeed, in a discussion about marketing it is very easy to get lost in any of a dozen different subject areas. There is nothing quite as frustrating as sitting down to discuss the overall marketing process with someone who immediately wants to argue the finer points of segmentation analysis, data mining, acquisition versus cross sell strategies, and market share measurements.

Each of these topics, and dozens more like it, are important parts of the marketing function. Much more important, though, is the need to develop an overall framework for marketing on which the telecommunications firm can count. Our first and most critical objective is to establish precisely this kind of framework. Remember, while most other industries have been practicing and integrating marketing into their organizations for decades, marketing in telecommunications has only recently become consequential. The majority of telecommunications firms approached the discipline of marketing as a byproduct of running their businesses. As a consequence, few are sure exactly how marketing should fit or how much attention to pay to the different activities that marketing people generate.

Before we can discuss strategies, marketing databases, and many of the other fine points, we should develop a basic understanding of how these fit together under the general marketing banner. In the following section, we take a general look at the overall marketing planning and execution process as it is practiced in most firms and industries. This process, developed over years of experience working in a number of different industries of varying sizes and types around the world, provides us with a good "best practices" perspective on how the marketing process could and in many cases should be run in a telecommunications firm.

Our objective here is *not* to promote any particular organizational scheme or to try to assign blame or responsibility for different aspects to various individuals or organizations. Our aim is to show you what a complete telecommunications marketing package would look like. The job and challenges of actually staffing and developing an organization to carry it out are, of course, a unique undertaking for each firm to consider.

Initiating the process

The first step is to develop an appreciation for how the marketing organization receives its instructions and sets its goals. This is actually more complicated than it might seem at first.

Sources of influence

Marketing deals with many sources of input. It is the nature of the marketing process that the marketing group is not set into action according to clear, discrete, measurable directives the way a corporate manufacturing or accounting group might be. Part of the job of the marketing organization is to take in and process dizzyingly complex inputs and to decide for the company how best to respond to them. We therefore refer to those agents that spur the marketing organization into action as *influences*.

Influence from outside the company (external)

We typically think of the marketing department as responding to external more than internal stimuli. Of all departments, marketing is the one most concerned with the activities of competitors and with overall marketplace trends and conditions. In fact, one could say that it is one of the primary jobs of marketing to keep the company informed as to those changing conditions and to be sure that the company is in a position to respond appropriately. Our list of external sources of influence includes the following:

1. *Competitive activities:* including campaigns and promotions;

2. *Industry trends and directions:* providing high-level insight into where the industry is headed;

3. *New technologies and innovations:* possibly having an impact on the company's long-term success;

4. *Regulation and government policies:* the telecommunications organization is heavily influenced in its decision making by the standards and policies established by governmental agencies.

Influence from inside the company (corporate)

Marketing is not only affected and driven by what is happening outside of the firm; it is equally concerned with the direction setting that comes to it from within the company. Major sources of influence in this area include the following:

5. *Corporate strategy and goal setting:* Instructions are provided by upper management regarding the definition of what the company's identity should be and what it wants to accomplish during the next year.

6. *Major corporate campaigns:* Major strategic promotional commitments are made by the company. These are the types of activities that shake the marketplace and require the concentrated effort of marketing and all other areas of the business. An example of a major corporate campaign of this nature would include MCI's "Friends and Family" campaign, which required the coordination of the activities of dozens of advertising and direct marketing campaigns and the focused activities of hundreds of employees to be successful.

7. *Budgetary constraints:* Limits on staff and expenditures set by the annual budget.

8. *Operational managers:* Managers of different operational units (e.g., product, sales, customer service managers) often bring specific requests or areas of investigation directly to marketing for consideration.

Influence from inside marketing (internal)

Also important to our understanding of the marketing process is some insight into the kinds of internal stimuli the marketing process creates, which will have an impact on the prioritization and execution of activities, including the following:

9. *Market research:* This group prepares surveys, focus groups, questionnaires, or other forms of externally generated information about customers, products, and company perception.

10. *Basic analysis:* Analysis includes development of studies into product and customer profitability, and competitive activities, usually performed by the marketing group's own staff. This analysis provides critical feedback about alternative solutions.

11. *Departmental objectives:* The department often operates according to its own set of internally generated policies and objectives in addition to all of these other sources.

The "forces" model of marketing influence

The three sources of influence over the marketer's decision-making process provide three different kinds of influence.

Motive force

Motivation is one way to influence the marketing organization into taking some action. Some of the motivating factors include corporate budgets and objectives, the actions of the competition, and the attitudes of customers. These influencers will tend to cause marketers to initiate certain activities and get the marketing process going.

Grounding force

Other influences are those that provide marketing with a clear idea about the direction in which they should be going, the constraints they need to deal with, and, in general, providing the road map for the chosen direction. Grounding activity usually comes from the regulatory environment, the telco's own network infrastructure, and its own operational efficiencies and idiosyncrasies.

Opposition force

Sometimes the same forces that provide the telco with a stable direction to follow can become so severe that they stop having a grounding effect and start being oppositional. Those same forces that can help the telco thrive in the marketplace can also prevent it from surviving. Ironically, regulation, network infrastructure, and the organizational structure can also exert a blocking force that prevents the telco from accomplishing its marketing objectives.

The spoiler force

Because the same forces can provide both grounding and oppositional force, we refer to the net impact of those factors as a *spoiler force*. Think of the spoiler on a race car, an accessory added to the back of a car that provides traction and steering capabilities, but can also slow down or speed the car's progress based on its angle and the way it is utilized. In the same way, these spoiler forces can be leveraged to either help or hurt the telco's marketing progress, depending on how they are used.

Applying forces to the marketing process

To help us to visualize these different operational forces and their impact, we can imagine the marketing process as a wheel. When it is functioning properly, a wheel rolls forward and moves the company in the right direction. When it is not working correctly, a wheel can be unmoving, stagnant, or erratic in its course.

As we can see in the diagram shown in Figure 2.1, the marketing process itself is propelled forward by the motive forces, and grounded or opposed by the spoiler forces (grounding and oppositional components). This image actually helps us understand a lot about the marketing process and its effectiveness in the business.

If the oppositional forces are too great, then the marketing process will be stalled. In this case, the company will need to change its strategy

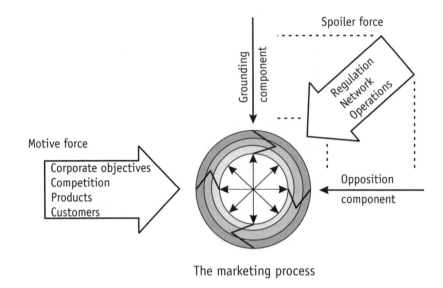

The marketing process

Figure 2.1 The marketing forces model.

and point itself in a different direction. If the company goes too far off the path established by the grounding force, then it will find itself in a market position that customers and regulators find unacceptable. If the company does not have a good handle on the management and focus of its motive forces, then the marketing process will not move forward at all.

Understanding campaigns

Given that it is the job of marketing to analyze all of the different inputs that these forces are exerting on the telco, and to turn it into some kind of positive forward-moving momentum, how does the process itself work? Although marketing performs many support functions, its main job is to create and execute campaigns.

Some future chapters are dedicated to the detailed understanding of campaigns. At this point, suffice it to say that a campaign is the organized execution of specific marketing activities to help communicate information about the telco to the customer and to motivate the customer in some way. You could say that the marketing process is what telcos formally go through to manage the delivery of campaigns.

The components of a campaign

Before we see how a campaign is put together, let's establish a basic definition of what a campaign is. In general, a marketing campaign is the organized delivery of a marketing message to a specific group of customers to accomplish some particular marketing objectives. To describe and execute a campaign, the marketing organization must define what we call the *five M's*:

- *Merchandise:* the product the telco will offer the customer;

- *Market:* the group of customers at whom the campaign is directed;

- *Message:* the message to be delivered;

- *Media:* how the message will be delivered;

- *Margin:* the price offered (optional).

To put together a campaign, then, a team must come up with ideas, rationalization, and often "models" that help justify selections of these variables in the pursuit of a particular set of corporate objectives.

Merchandise—what do telcos sell?

We could easily list a dozen different products and groups of product offerings that make up the inventory of "merchandise" that the telco has to offer customers. For simplicity's sake, however, we will concentrate on only the three major product groups.

- *Long distance:* delivery of telephone messages from one location to another, usually across state, territory, or national boundaries;

- *Wireless:* includes cellular, PCS, satellite, and other forms of non-wired communications;

- *Wireline:* competitive local exchange carriers (CLECs) still represent an incredibly large portion of the overall telco business. (They are the companies that own the actual connection of the phone to the consumer's home or place of business.)

Although every telco has a different combination of products and services they offer, these three represent the lion's share of what the telco business is all about. Looking at them will help us understand how the differences between them affect the marketing process.

The market—who are the telco customers?

Not only is understanding what we are trying to sell important to the marketer, but so is figuring out to whom we want to sell. The art and science of defining the targets for campaigns could be the subject of many books and we will discuss the subjects of segmentation and modeling in greater detail in later chapters. In general, however, three major market groups form the customer base for telcos.

To fully understand the process of campaign development, we must add this dimension to our model as well. Included in these target groups are the *consumer market*, the *corporate* (large business) *market*, and the *small/medium business market*.

The consumer market

For most people thinking about telecommunications and marketing, the consumer market comes to mind: individuals, homeowners, and apartment dwellers. This market is a large one.

The corporate market

Although the consumer market certainly represents the largest market for telcos in terms of the number of subscribers, it is by no means the largest market for the amount of services consumed. Typical consumers may use a large amount of long-distance, wireless, and wireline service in the aggregate, but clearly corporations consume hundreds of times as much as a group of citizens in any given time.

The small business market

Somewhere between the individual consumer and the giant corporation is another very large market consisting of the small and medium-sized businesses who have buying behaviors and usage patterns that represent a hybrid of the other two.

Defining messages: the major objectives for campaign activities

When we consider the incredible diversity and confusion that typifies the different inputs of information that marketers must deal with, we might wonder how they go about making any decisions at all. "These customers have been buying this much." "These competitors have dropped their prices by twenty percent." "These product managers are saying that they need more marketing support for their areas." Marketers have to take such statements and turn them into specific plans of action.

After considering many different factors, the marketer may decide that the best use of the organization's resources is to pursue a limited set of specific objectives and employ one of three basic approaches. Each of these approaches can help set the direction and define how the content for the messages to be delivered will be established.

The core approaches the marketing department can use to improve the profitability, revenue, or market share of the firm, and the key objectives they set for these activities, are as follows:

- *Acquisition:* turning noncustomers into customers;

- *Wallet-share enhancement:* getting existing customers to spend more money or generate more profit;

- *Retention:* convincing existing customers not switch to the competition.

The marketers may be told "We want to increase the revenue for the sales of wireless services," or "Our objective is to become the largest provider of long-distance services in this marketplace," or "Our profit margins are dangerously low." In any case, the analysis will ultimately lead them to the decision to try to effect the desired change through one of the three approaches, the acquisition, wallet-share, and retention processes.

Different media and their impact on campaigns

The decision-making process involved in the development and execution of campaigns for merchandise, market, and message criteria is consistent for all types of campaigns. There are, of course, variations depending on the situation, but the process itself is very similar in all cases.

Decisions made about which media to use, however, have significant impacts on the ability of the marketing organization to manage itself. Ultimately, when the rest of the background work is finished, the truly exciting and interesting part of the marketing process has to do with the construction of different messages to send to the customer and the different media that can be used to deliver them.

Unfortunately, for the rest of the business world, this exciting process is also incredibly creative, free floating, and therefore "unmanageable" in many ways. That makes the job of building marketing support systems very challenging. However, underneath the creative frenzy and artistic expression that typify this aspect of the business are some basic core processes that can be understood and managed. We find that, embedded in the marketing process, are three distinct subprocesses (variations on the main process) according to the advertising media with which the marketer is working. These media include:

- *Mass media*. Included in the mass media category are those advertising activities that involve television, radio, and print. This use of mass media is the most expensive, most powerful, and most difficult to manage of the three media types.

- *Direct media*. Secondary to the mass media approach are the advertising techniques known as *direct marketing*. These include phone solicitation (inbound and outbound) and sending catalogs and other forms of advertisements to individuals in their home or office. Although not as popular as mass marketing with many telecommunications firms, the direct marketing approach is quickly gaining in importance as the benefits of greater control and more focused activity become apparent.

- *Promotional activity*. The last type of marketing activity to consider is the promotional activity area. This includes participation by the telco in trade shows, community activities, and a variety of other "public relations" types of activity.

Each of these delivery media has some common characteristics, and the actual management of the process of delivery is incredibly diverse. They are a special challenge for us to consider as we continue to define the marketing process.

Margins: the pricing and profitability issue

Although it would be impossible to put together a campaign without making some specific decisions about the merchandise, market, message, and media involved in its execution, the price and profitability information, though important, is hardly ever explicitly included. There are two reasons for this:

1. The regulatory environment, in terms of how much leeway is allowed when pricing products, limits many telcos.

2. Most telcos are not sure how profitable their products are or what they should optimally charge for them.

In the majority of cases, the margin characteristics of a campaign are simply assumed or ignored. That, however, does not mean that they are not important.

Multidimensional view of telco marketing

To understand the process of marketing in a telecommunications firm, the number of different forces that work on it, and the forms that the marketing process can take depending on these variables, we need to keep all of these different dimensions in mind. Table 2.1 shows the five dimensions of telco marketing.

Line of business combinations

Each of the different dimensions of the marketing process is important by itself. When we see all the dimensions together in action, we realize the scope and breadth of the marketing process.

Table 2.1
The Five Dimensions of Telco Marketing and of a Telco Marketing Campaign

Merchandise (lines of business)	Market	Messages (objective)	Media	Margin
Long distance Wireless Wireline (local)	Consumer Corporate Small/medium business	Acquisition Wallet share Retention	Mass Direct Promotion	Price and profitability

Long-distance combinations

Looking at the synergy between these five dimensions in the long-distance arena, we find some of the following:

- *Market.* Long-distance companies have spent considerable time and energy concentrating on all three aspects of their market. In this particular line of business, different companies tend to specialize in meeting the needs of one type of customers. Some long-distance telcos have developed a reputation as *corporate* long-distance providers and others are becoming known as *consumer*-based telcos. In reality, of course, they pursue all markets actively, but to differing levels of success.

- *Margin.* In general, long-distance carriers tend to be extremely profitable and well positioned.

- *Message and media.* Some of the most famous and interesting of all the advertising campaigns in telecommunications are seen in the long-distance arena. Long-distance telcos do *a lot* of mass marketing (since they have good margins, the higher costs are not an issue) and shift their messages depending on market position and maturity. In newly deregulated markets, acquisition programs are critical. As the markets mature, retention seems to be the key. In general, the long-distance telco will not be too involved in cross-sells unless they have acquired other lines of business along with their own.

Wireless combinations

The wireless industry is the newest and most volatile of the telco lines of business at this time. Since the products are new and the regulatory environment less restrictive and since a wireless vendor is not bound by physical location, marketing in this area has been the most aggressive and productive.

- *Market and margin.* A significant shift is occurring in the wireless sector as it matures. The industry started as an almost exclusively corporate product, which gradually drifted into the small and medium business space. Today, most wireless telcos find that a growing number of their customers are consumers. Of course, because consumers do not use the phone nearly as much as corporations, the revenue per customer has shrunk in the process.

- *Media and messages.* Wireless companies, which tend to be smaller in size and finance as well as more limited in their geographical scope, have come to rely heavily on the local mass media (newspaper and radio) and on direct marketing, much more so than the long-distance companies. This is encouraged further by the reduced margins that each customer is bringing to them. In terms of messages, a new wireless telco will follow an almost predictable pattern. The company starts with an aggressive acquisition campaign to accumulate the subscriber base, which is followed by a frenzy of retention activities as competitors begin to erode the profitable customer base.

Wireline combinations

The last line of business to consider is the wireline (local wire) service. In most cases, these are still the same old telcos that before deregulation represented the entire phone company.

- *Market.* These organizations are well established in all three of the major markets (consumer, small business, and corporate) and despite regulatory attempts they tend to be extremely well entrenched in at least the consumer and small business arenas. Highly profitable and geographically clustered corporate sectors are causing a great deal of competition in that area.

- *Margin and message.* Margins for these telcos, although not the best they have ever been, are strong and improving quite radically, especially in those cases where the organizations concentrate their activities on cross-sell. Since these companies still have a relatively captive market, at least for the time being, they can concentrate on adding more and more products to their portfolio and convincing consumers to add on to their existing service (ISDN, T1, Internet, pager, and other add-on services).

- *Media.* These organizations make use of mass media and, because of the nature of their market (known and captive), they rely most heavily on direct marketing.

3

The Core Marketing Process

What is essential is often hidden to the eye.
Antoine De Saint-Exupery

There are good apples and bad apples, but the worst are the apples that are rotten at the core. They look wonderful on the outside, but reveal their decay only after we have taken a bite of them.
Anonymous

So far, in our exploration of the marketing process in the telecommunications firm, we have established that it is the job of marketing to deal with the many forces that work on the telco to move and change in different directions. We have seen that the marketing process itself is one of creating and delivering campaigns that help make all of this happen. We determined that campaigns can be best understood as being formulated on five dimensional levels (merchandise, market, media, message, margin). Now we need to describe how this marketing process itself works.

The elusive marketing process

Although marketing is a familiar concept and most organizations are involved in it, once you begin to analyze this notion it becomes very difficult to determine just what that process is. Frankly, marketing is just too hectic, creative, and disorganized to expect the marketers to worry about what process they are following. That, of course, makes it difficult for a person unfamiliar with marketing to understand what is going on most of the time. More importantly, because the vast majority of telecommunications firms has little or no formal marketing history or processes to rely on, the marketing landscape can be especially confused and conflicted.

The core marketing process

While we would be hard pressed to find a definition of a formal, well-documented, highly structured marketing process, by looking at how the majority of telcos functions, it is possible to identify the underlying processes that drive the organization. These processes are often so subtle and assumed that everyone takes them for granted. We refer to them, therefore, as *core processes.* Just as you don't see the apple core until you take a bite, so is this process hidden beneath the exterior that marketers present. The other reason we call them core processes is that they are critical to the lifeblood of the marketing organization.

So far, we have established the following:

1. The primary activity of a telco marketing organization is to create and execute campaigns.

2. The campaigns are created according to the information that marketers gather from the various forces that exert influence over the firm and provide insight into the needs of customers.

3. Based on that input, the marketing organization identifies the optimum arrangement of market, message, medium, and margin decisions so that the campaigns will most effectively respond to those forces.

Now let's see how the marketing process itself is organized. Basically, a typical marketing organization's activities can be divided into the following major steps or phases:

1. *Prioritization and goal setting:* the process of deciding what to focus on and to build teams that pursue those objectives;

2. *Modeling:* the process of performing segmentation, scoring, and other kinds of analyses that help marketing decide on which campaigns to develop;

3. *Campaign development:* the process of choosing, creating, and preparing the advertising medium and the message to be delivered to the customers as well as running tests to validate the marketers assumptions;

4. *Campaign execution:* the process of actually sending the messages through the respective media.

Marketing and the execution of the different phases of the process are in constant motion, always changing, and continually reevaluating. The process is not serial or linear and has no true beginning or end. To reflect this movement, we represent the different phases as parts of a wheel (Figure 3.1). Because each phase of the process is dependent on feedback and analytical information from the other phases at all times, we let the spokes represent this continuous cross-pollination of information and data between the different phases. The diagram of Figure 3.1 eloquently demonstrates the relationships between the phases of the marketing process, each phase feeding into the next phase and simultaneously providing information to all the other phases. Given this basic framework, study each of the phases or steps that the marketer will go through in the process of creating and executing campaigns.

Prioritization and goal setting

The marketing process has to deal with many different sources of input, feedback, and impetus. In fact, in most telecommunications firms there is so much input from so many different sources and with so many

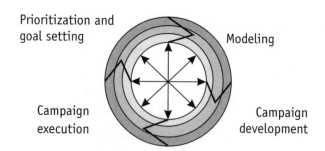

Prioritization and goal setting

Modeling

Campaign execution

Campaign development

Figure 3.1 The core marketing processes.

contradictory objectives that control over the process can very quickly be lost. What keeps this from happening is what we call the *prioritization and goal-setting* function.

This process may be extremely dysfunctional in some organizations, nonexistent in others, and well run in a small percentage of cases. Most of the major challenges to the execution of the marketing process occur here, in the area of prioritization and goal setting.

Stages of prioritization and goal setting

During this initial step of the process, the major components of a marketing plan are assembled. These steps involve identification of sponsors, creation of projects, assembly of teams, identification of objectives, and identification of constraints.

Identifying sponsors

Every marketing activity must have a sponsor, someone who believes that marketing should be doing something about some situation. While outside influences such as competition or technological innovation may be what inspires marketing to investigate different possibilities, it is the sponsor who actually makes that kind of pursuit a reality.

Creating projects

Once a sponsor has decided that something needs to be done, this person must formally describe the problem or objective and arrange for the financing. The identity of the sponsor and the importance of the project will often dictate the extent of the formality. If the sponsor is a manager in the marketing department itself, then the creation of a particular project may involve no more formality than sending a memo to a marketing staff member to investigate further. If the sponsor is an operational manager, the project creation may be part of a much larger operational activity. If the sponsor is part of upper management, the project may have funding allocated that will be spent with an advertising agency.

Assembling a team
Next, the sponsor will formulate a team of people to investigate the issues. These teams will be made up of people from the sponsor's organization, from marketing, from computer systems, and from outside firms.

Identifying objectives
One of the team's first jobs is to formally define the objectives of the project and to translate those objectives into tangible terms. Typically, marketing project objectives are stated in terms of *product line* (the service or product involved), *market segment* (the customers to be approached), and *measurable metric* (revenue, profit, or market share). Sponsors typically have only a general notion of what the project team is to accomplish and it is the team's responsibility to formalize that objective so that it can be executed effectively.

Identifying constraints
Another important job of the project team is to identify the project's constraints. Budget, regulatory, and capacity constraints affect the range of solutions the team can employ.

Modeling
Once the project team has been assembled and financed and the objectives have been clearly stated, the team moves from prioritization and goal-setting mode into *modeling* mode. A model is a proposal of the method that will accomplish the sponsor's objectives and a mathematical proof that shows the proposed solution to be reasonable and viable. The goal of modeling, then, is to develop a plan, or a series of plans, that will accomplish the goals. The modeling process is the process where most of

the analytics disciplines are employed. (Some analytics are also involved in the testing and development of campaigns and in the postmortem analysis of completed campaigns.)

Definitions of key analytics

Many different kinds and forms of marketing analytical work can be employed. Some of the key categories are discussed in the following subsections.

Segmentation

Segmentation is the most common form of marketing analysis work. Its main function is to critically analyze customers, their characteristics, and their behaviors to help the marketer understand them and predict their actions. Segmentation is the number one process that occurs in the modeling phase.

Scoring

Scoring is the process of ranking and/or categorizing customers according to how likely they are to behave in a certain way or how high or low they rank with all other customers for given criteria. Scoring is used extensively in the execution of direct marketing activities. It can also be utilized to assign overall value to a customer (the customer value function), or to appraise their viability in terms of credit risk (credit scoring) or fraud risk (fraud scoring).

Product and profitability analysis

Whereas segmentation helps the marketer to understand groups of customers better, *product and profitability* analysis provides insight into the nature of the products and services the company provides.

Compound and simple modeling

In the modeling process, the objective is to create discrete views of customers/prospects that help define how effective a given campaign will ultimately be. There are many kinds of models, however, and that often causes confusion. One way we divide the different types of models is to separate them into simple and compound categories.

We use the term *simple model* to refer to any discrete, autonomous, mathematical proof that defines one particular aspect of a customer's or product's behavior or any other kind of discrete observation. We use the term *compound model* to describe those models that are actually combinations of simple models. For example, suppose I build a simple model that tells me how likely customers of different age groups are to respond to an ad. Then, suppose I build another simple model that tells me how likely customers of different income levels are to respond to the same ad. If I combine these two simple models, the larger compound model will provide me with much more information than either model can separately.

The modeler

For the nonmodeler, one of the most elusive and frustrating things about the modeling process is trying to keep track of the modelers as they go through the many steps in the process. The modeling process is extremely complex, recursive, and creative and modelers need to deal with dozens of objectives, hundreds of constraints, and thousands of variables when setting out to develop an optimum model for a given situation. Even more importantly, the effectiveness of the models generated depends in no small part on the business acumen and statistical sophistication of the person running the models. It should come as no surprise, then, that the role of the modeler in the telecommunications marketing organization is one of the most critical and valued and one of the toughest to fill.

Campaign development

Once the modeling process has been completed and the decision has been made about which model to use, the job of marketing shifts from the modeler to the campaign developer. Campaign developers turn the models into specific, executable plans. A number of processes are involved in campaign development:

1. *Media and message finalization.* Reach a decision on the specific message to use and how it will be delivered.

2. *Media selection and negotiation.* Choose specific organizations and individuals to deliver the messages and negotiate the cost of that delivery. (This varies, obviously, depending on the medium selected—radio, television, direct mail, telemarketing, and so on.) Media selection is often influenced strongly by the time frame involved, the message, or the target group.

3. *Prospect targeting and list scoring.* In the case of direct mail and telemarketing, one of the main jobs of the campaign developer is to select the list of prospects and to determine which specific people on that list will be targeted (the scoring process).

Campaign development is based on the three major objectives of acquisition, wallet-share enhancement, or retention just as the modeling activities are.

Campaign execution

Once the campaigns for a given time period (monthly, quarterly, or annually) have been developed, the marketer will then be responsible for setting them in motion. Campaign execution is what marketing is all about. It may involve nothing more than making a phone call to the advertising agency or may be as complicated as initiating and monitoring the day-to-day activities of a call center or mail room operation. The work that needs to be done to successfully execute a direct marketing campaign can be very extensive. Ultimately, however, the message is delivered and the company is ready to analyze the results.

Feedback mechanisms and the marketing database

The analysis performed by marketers in each of the discrete parts of this process would be of limited value, if it were simply executed, used, and discarded. If there were no way to capture the knowledge discovered during one part of the process and share it with those in another phase, then the process of marketing would be much more expensive and much less effective.

Post-campaign analysis

The key to effective and efficient marketing is to put into place a mechanism that tracks how well a campaign is doing as it progresses. The ability to report on how close the organization comes to meeting the objectives set up at the beginning of the campaign provides valuable information about how campaigns might be made more effective in the future. Post-campaign analysis is an integral part of the overall marketing life cycle and feeds information right back into the initial steps (product, customer, and competitive analysis), which then allows the company to reevaluate its strategy and tactics and start the process all over again.

General feedback throughout the process

In addition to the specific, detailed, and formal process of diagnosing the results of specific campaigns, most marketing organizations also build in extensive, dynamic, real-time feedback mechanisms between all of the different processes. In this way, everyone in the marketing organization is able to benefit from the activities of everyone else. In general, each of the different marketing processes that can be managed also produces feedback information critical to running the marketing organization.

Providing support for the marketing process

The marketing process is an extremely complex and analytical one. In fact, the amount of number crunching involved has become so intense that many organizations have built special systems dedicated to managing and coordinating all of these different analytical and operational proceedings.

To complete our model of the overall marketing process, therefore, we need also to take into account and define the underlying infrastructures used to execute the marketing processes. This foundation, like the marketing process itself, is complex. In fact, it involves three different layers of support including (1) specialized processes, procedures, and techniques; (2) application of specialized skills; and (3) core computer technology support in the form of systems, tools, and collections of data. See Figure 3.2.

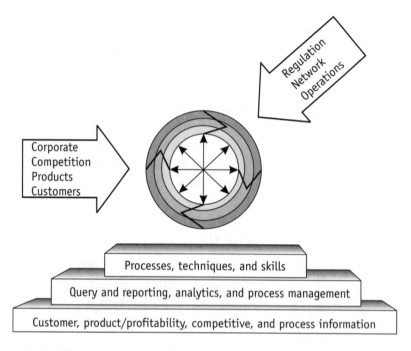

Figure 3.2   The complete marketing process.

Processes, techniques, and skills

Each of the four core processes that make up the marketing process has certain assumptions associated with it about the skills, techniques, and procedures necessary to make them work. In the prioritization and goal-setting phase, for example, the ability to visualize complex problems, to communicate effectively, and to create and motivate teams are crucial to success. During the modeling phase, analytical skills, especially those associated with statistical analysis and abstract interpretation of facts and figures, are critical. During campaign development, intimate knowledge of the nature of the different media and the crafting of messages as an art form cannot be replaced. Finally, during campaign execution, administrative and management abilities are the skills that define success.

Despite the best skills and techniques in the world, however, the marketing organization will not be able to hold all the diverse pieces of the marketing puzzle together very well unless they have a good computer system that makes the whole thing work. As the world of marketing continues to increase its depth and breadth of operation, the need for a single, comprehensive source to manage all of its processes continues to grow. As we shall see throughout the rest of this book, the marketing database is quickly establishing itself as the key component in the ability of the telco to manage marketing efforts successfully.

The telco marketing database

To assist the marketer in the execution of the many different functions involved, the information technology (IT) organization often creates specialized marketing database environments to support them. These systems can be large, integrated environments, or consist of little more than a haphazard assortment of disconnected programs, applications, and data stores. As the telco grows, however, and as the competitive climate continues to become more customer-centric, these systems tend to become more and more structured.

Three primary functions delivered by a telco marketing database

Although there is an incredibly diverse assortment of systems that come into play for marketing support, they generally fall under three major categories.

Query and reporting

Query and reporting systems are the least "interesting" or "glamorous" of the marketing database support systems, but they usually do most of the real work in the environment. These systems allow people inside and outside of marketing to view and analyze information about all aspects of the marketing process and make use of all sorts of different customer, product, and profitability perspectives. They are usually supported by one or more query, reporting, or online analytical processing (OLAP) systems that provide easy access to all data for all users.

Analytics

While query and reporting systems are the general-purpose tools that everyone uses, the analytics applications are the systems that the serious statistician uses to do detailed segmentation, forecast, regression, and other kinds of studies. The analytics environment usually consists of one or more tools, typically data mining and statistical analysis packages, and some very sophisticated users who know how to apply them.

Process management

Sometimes, the particular operation is so large or critical to the overall marketing success that the company actually invests in a formal transaction processing type system to manage it. In most cases, however, the process is managed in less structured ways. At its worst, the process is allowed to run without any structure at all. A few of these processes are discussed next.

Project management   Project management is the process of keeping track of the various projects that different sponsors are running in the marketing area. It is usually one of the least sophisticated and least often utilized management functions in marketing.

Campaign management   Campaign management revolves around keeping track of the different campaigns that an organization runs, the people they are directed toward, and the diverse impacts they have. Formal campaign management systems are most common in organizations that run many direct marketing campaigns.

List management   List management is the process of monitoring the lists of customer and prospect names and making sure that they are acquired, prepared, and utilized efficiently. List management systems are most often found in organizations that buy much external data for running their direct marketing efforts.

Customer management   Customer management systems attempt to centrally store and manage all information about a customer. Often called *customer information systems,* or CISs, they are many times spun off from the billing system and will often be under the control of some organization other than marketing.

Contact management   Contact management systems are a special type of customer management system. They not only centrally maintain information about the customer, but actually keep track of every single contact any one employee has with that customer. These systems, usually run by the customer service organization, attempt to maximize the relationship between customer and company through better tracking of information about messages sent and responded to on all levels.

Segment management An area receiving a lot more attention recently is segment management. This process manages the different segmentation categorizations to which a customer is assigned and measures how effective each is in the attainment of marketing objectives.

Product management Product profitability and management systems, once the property of only the largest retailers and manufacturers, are gaining in popularity as an adjunct to the tools available to the telecommunications marketing firm. The ability to track and report on the profitability, capacity, and applicability of different products provides a key input into the analytical job of the marketer.

Model management Finally, the entire process of model building itself is one that also needs to be managed.

Core data collections used to support marketing

To make all of these different marketing processes work, the computer systems department is responsible for providing information that is used to support each of the marketing processes (prioritization, modeling, campaign planning, and campaign execution) through each of the support systems (query and reporting, analytics, and process management). The incredible power that the marketing discipline is able to exercise over the profits and efficiencies of a company are made possible through the extensive use of this information.

Customer information

This information allows the marketers to work on the development of alternative strategies for viewing and approaching the customer marketplace. They develop different plans, called *segmentation schemes,* for categorization of customers. With these schemes the analysts can develop an understanding of what customer belongs to which segments and how they will behave. The objective of segmentation analysis, like product analysis, is to determine how to maximize profits through the selective management of different groups of customers. Customer information comes from internal and external sources and is the principal collection of data that marketing uses to make decisions.

Product and profitability information

Using this information, analysts look at the different products and services that the telecommunications company is offering to its customers to determine exactly how profitable they are. This analysis includes profitability, pricing, and costing information. Without a good understanding of how profitable products are, it is impossible for the marketing strategist to develop a profit maximization plan. Although this area is one of the least developed for most telcos, it is usually the most critical to good decision making and is increasing in importance with each new generation of system.

Competitive information

To do a good job of planning, the analysts also need information about the competition and the industry overall. To that end, analysts will look, for example, at how well the competition is doing in each of the markets in which the company has decided to compete. A close examination is also given to how well the industry is performing overall and how young or mature each of the different product offerings are in terms of the overall product life cycle.

Process information

Finally, there are data stores that keep track of the different activities that marketing itself has generated. The amount of marketing process information available, of course, is directly related to the amount of marketing process automation that is being handled by the system.

Timing issues with the marketing process

The marketing planning and execution process that we have discussed describes how marketing is done in a typical corporate environment. It would be a mistake, however, to assume that, since the steps involved are presented in a logical, sequential order, the marketing process is also bound to this same kind of rigid scheduling. On the contrary, the marketing practices in most organizations require that the marketers be able to pick up and execute different parts of this process at different points of time, depending on whatever requires the most immediate attention.

In the ideal world, a marketer might like to organize the world in such a way that in the month of January they do product analysis, February is spent with customer analysis, and March is dedicated to the development of competitive analysis. The marketing department could then spend April putting the strategy together, organize campaigns in May, and then dedicate the rest of the year to running the different marketing campaigns.

Of course, in reality the marketer ends up needing to do all of these things at the same time while making sure the right kind of analysis is used to help support the right kind of strategy decisions. The typical marketing professional is required to juggle all of the issues and procedures and to understand how each piece fits with all the other pieces. See Figure 3.3.

Figure 3.3   The marketer's juggling act.

Part 2

Understanding Marketing Campaigns

4

Telecommunications Strategy and Campaigns

A man to carry on a successful business must have imagination. He must see things as in a vision, a dream of the whole thing.

Charles M. Schwab

Strategy and campaigning

Thus far, we have limited our discussions of the marketing process to the rather tedious and mundane planning and overview perspective. This dry, computational, and factually based reality is only a small part of the world where the marketing person lives, however. The other half, the one that we will address now, is a world defined by bold visions, high risks, high rewards, and a great deal of artistic investment and business expertise. We refer, of course, to the area of marketing strategy development and execution of specific marketing campaigns.

Definition of terms

Before we get too far into the discussion of strategy, let's take a moment to define some of the terminology. The term *strategy* can have a number of different meanings and we want to be absolutely clear on what it means to us in this case.

Webster's Dictionary defines strategy in two ways: (1) the science or art of planning and directing large-scale ... movements and operations, and (2) a plan or method for achieving a specific goal. We will, throughout the course of this book, make use of this term in ways that include both of these meanings. When it becomes important, we will use the term *corporate strategy* to refer to the science or art of planning and directing large-scale operations. When talking about specific, tactical plans or methods for achieving a goal, we will use the simple term *strategy*.

The marketing department, therefore, helps put together a *corporate strategy*, which is then executed through construction and execution of a number of specific *strategies*, as expressed in the form of marketing campaigns. Each campaign, then, represents one of the *strategies* that we will use to help meet the objectives of the overall *corporate strategy*.

Strategy in the world of telecommunications

To participate in the development of strategy for a telecommunications firm we need to understand a little more about exactly how that business works and what the real nature of strategic decision making is all about. Of course, we are not able to provide a completely thorough explanation of all the subtleties of corporate strategy at this level. What we can do, however, is make note of some of the more obvious and pressing "hot buttons" that make the development of telecommunications strategy so interesting and challenging.

Principal assets of a telecommunications company

The development of any strategy must include, first and foremost, an evaluation of the company's assets and an exploration of how these assets can best be utilized to competitive advantage. In simplistic terms, we can say that the telecommunications company has three principal assets (Figure 4.1).

1. *Organizational assets:* the people and things that make the business work;

2. *Marketing assets:* a collection of customers, consumer perceptions, and attitudes about the company;

3. *Network assets:* the heart and soul that defines the telecommunications company.

We will consider each of these in more detail—what they include, how they work together, and how they fit into the overall strategy development process.

Organizational assets

The easiest to recognize capital of any company is its organizational assets, which is everything that drives the day-to-day running of the business—human, procedural, and physical. Included in organizational assets are the company's buildings, offices, and other support facilities, as well as all the employees and the years of accumulated knowledge and expertise they maintain. Other important assets in this area include the many processes, policies, and procedures developed over the years to

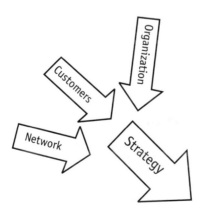

Figure 4.1   The three components of a telecommunications strategy.

create a large, well-run corporate machine. Critical components of any company's organizational infrastructure are, of course, the computer-based systems that make the company work.

The strategic use of the organizational assets

The discipline known as *business management science* is based on organizational asset management. In this sense, the telecommunications firm is no different than any other business. Business process reengineering, management control and administration, downsizing, reorganization, budgeting, and organizational goal setting are well-known ways to improve the company's profitability. These strategies are ways of tuning the organizational assets and are used in every industry to make the business run more efficiently and more profitably.

The telecommunications value chain

Recent advances in the study of management science have resulted in the development of a discipline known as *value chain analysis*. According to this approach, the business is best understood as a series of major functional components or links in the value chain. Each of these links serves to identify a different major function that must be performed within the business. Most telecommunications firms have value chains that include the following tasks:

- Research and development (new product development and network support);

- Acquisition (securing the right to do business in a given area);

- Network infrastructure planning and development (planning and building the network that will carry traffic);

- Network maintenance and support (the process of keeping that network functioning);

- Provisioning (setting up and breaking down service capabilities);

- Activation (turning customer service on/off);

- Service order processing (managing customer requests);

- Billing (tracking and billing customers);

- Marketing (managing the relationship with the customer);

- Customer service (keeping customers happy);

- Sales (establishing and maintaining customer relationships);

- Finance and accounting (keeping track of costs and revenues);

- Credit management (managing the firm's credit risk);

- Operations (network and business);

- Information and computer technology (computer systems used to run the business).

According to advocates of this approach, the optimum use of the organizational assets is reached when the company's value chain is understood and tuned appropriately. We will, in later chapters, take a closer look at the telecommunications value chain and see how its manipulation can aid in the development of a marketing strategy. At this point, it is enough to realize that the best marketing and network in the world will do a telco little good if the business organization itself is not sound and efficient. Organizational assets, therefore, play a critical role in the development of any corporate strategy.

Marketing assets (customers and market image)

The second collection of assets critical to the development of a corporate strategy is the telecommunications firm's marketing-related capital. Included in this category are:

- The customers that the company currently provides service to;

- The history of services that former and current customers and the marketplace in general are aware of;

- The history of performance, efficiency, and revenue generation that government agencies and the investment community are aware of.

These assets, both tangible and intangible, provide the marketing strategist with the starting point and the building blocks on which the new marketing strategy will be based.

The greatest assets: customers and customer history

Companies are beginning to realize that their survival in this increasingly competitive market is dependent in no small measure on how well they manage the customers. In every industry and in every geographic marketplace around the world, companies are finding that their customers are the most important assets of all.

As more and more firms have learned the art and science of managing their organizational assets, it is becoming increasingly difficult for a company to truly distinguish itself as a good, low-cost provider of goods and/or services. Customers have come to take quality of service as a given in most areas of life. Consequently, they are depending more and more on their impressions and memories of a company in their buying decisions. The company that takes care of its customers and provides them with a solid history of positive experiences and impressions is the one that will keep those customers over the long run.

Branding and loyalty

Studies have shown repeatedly that customers will buy goods for a variety of reasons, but that one of the most important is their familiarity with whom they are buying from. During the past several decades, manufacturing companies have done much research into this phenomenon and have developed the concept of *branding*. Branding is the process of establishing an image in the consumer's mind of quality, dependability, and a positive experience in general whenever they buy or use the product or service in question.

A strong brand name is an incredibly powerful and lucrative asset for a company and the numbers back up the claim. Branding studies have shown that products with a strong brand image not only tend to dominate the market in terms of sales volume (quantities sold), but also claim the best profit margins (branded products tend to sell at a higher price).

Branding also elicits a certain level of loyalty. Customers who associate positively with a strong brand name tend to stay loyal to that brand despite sale discounts and other incentives that competitors may offer.

Branding and the telecommunications firm

Establishing a strong corporate image and creating a brand image in the minds of consumers is one of the major goals of every telecommunications marketer. It is the objective for many of the strategies that are being put together.

Of course, establishing a brand name is not easy in any industry and it is especially difficult in telecommunications. The history of the telecommunications industry itself conspires to undermine the branding strategies of many firms.

The role of the marketing database

Key to the management of any marketing operation, and especially important to the telecommunications firm, is the ability to measure, analyze, and plan marketing activities on the basis of good solid factual information. Most telecommunications firms are ill equipped to perform this kind of analysis, at least at the outset. After all, they were able to survive without this kind of analysis and detailed computation for decades.

What they need is an operational environment where the marketers can do this kind of work quickly and efficiently. More and more companies are building special kinds of data warehouses known as *marketing databases* to support these activities. A marketing database is a system that attempts to gather all of the different aspects of marketing planning and execution under the control of one centralized environment. When a marketing database is built well, it very quickly turns into one of the company's most valuable marketing assets.

The last part of this book, Part 5, is dedicated to a more thorough exploration of the role and functionality of a telecommunications marketing database.

Customer relationship management

A relatively new discipline known as *customer relationship management* (CRM) is available to keep track of customer relations. It is a cross-disciplinary approach that attempts to fuse the activities of marketing, customer service, provisioning, sales, and billing into one cohesive customer relationship management environment.

The CRM initiative is usually heralded by the creation of a flagship *customer relationship management system* (CRMS) that ties each of these business areas together and coordinates their activities on a customer-by-customer basis. While CRMSs are certainly consistent with the goals of marketing and the establishment of marketing databases, they are still relatively new and untested. It is still not known whether the approach will be economically effective. When eventually they do work, however, they will most certainly become an important part of the company's marketing asset base.

Network assets

Every business has organizational and marketing assets. They may be unique to a given company, industry, geographic location, or history, but the disciplines that drive them are the same for a retailer, manufacturer, airline, or telecommunications firm. What makes the telecommunications firm unique, however, is its most important and principal asset—its network. Manufacturers have factories, banks have money, and telcos have networks. The network is what separates the telco from its competition and what defines how much it can do and how well.

The business of building and supporting network infrastructures —never an easy task—has become incredibly difficult during the past few years. If you remember from our earlier discussions, the main reason that governments established telcos as monopolies in the first place was because of the enormous investments in property, switches, wires, and other assets necessary to make a fully functioning telecommunications network a reality.

So, keeping in mind that a network is extremely expensive to build, it follows that the telco would want to be relatively sure that a very large number of people would be using it in a relatively short amount of time. Imagine how long such a company would last if it put up hundreds of miles of wiring to locations where there are no people! (Don't laugh, it has happened.)

One other critical point about networks is that telecommunications technology is changing so quickly that it is impossible to keep the network up to date with the latest innovations.

Network investment: the number one strategy component

While the operational infrastructure and the marketing-based assets (customers and market image) are certainly important to the telco, neither means anything if the right network investment and development strategy is not in place. In fact, investment in networks is exactly what the telecommunications industry is about today.

The number one target of investment money around the world

For the past several years now, the stock market, investment banking companies, private investors, governments, and other businesses have all been investing more and more in telecommunications companies. Almost every substantial and reputable source of information about investment and money trends reports that telcos are receiving the lion's share of the investment money that goes out these days. More money is invested in telcos than in computer companies, airlines, manufacturers, retailers, or any other group.

The number one IPO stock in the world

Equally pronounced is the less dominant representation of telcos in the initial public offering (IPO) arena. (An IPO is the event when a firm releases company stock for trading on the open stock market for the first time.) Telecommunications firms show up repeatedly as the most often released and best received IPOs.

International investment reaches all time high

Equally pronounced is the way in which different governments and telecommunications companies are investing in each other. As more telcos become deregulated, more companies are stepping in to participate. Major telecommunications companies such as AT&T, Sprint, MCI, British Telecom, Nippon Telephone and Telegraph, Telefonica Espana, and a host of others are racing to form partnerships with smaller national telcos or to bid for the rights to compete in already established markets. A lot of money has already been poured into these markets and all indications are that the trend will continue for some time yet.

Why is so much investment needed?

Why are so many companies spending so much money on the development of newer, faster, and better telecommunications networks? One would think that all of the phone lines that are needed are already in place and that just a little bit of upgrading here and there would take care of whatever people's needs might be. Most companies, governments, industry observers, and telcos, however, do not agree with this view. The way they see it is that we are undergoing a major revolution, not only in the way we use telecommunications services, but in the very way we do business. The telecommunications industry is leading the charge into a new world of commerce that depends on the sophistication and capacity of its telecommunications networks to survive and thrive.

Economic analysts have noted a very interesting trend in the economic viability of different geographic areas of the world. They are finding that the telecommunications capacity of a country or region is one of the most indicative measures of how well that area will do. Businesses will not locate where good, high-capacity, high-quality telecommunications channels are not in place. They cannot do business with partners when they cannot interface with them on a myriad of different telecommunications levels. Their employees cannot function efficiently and effectively if the telecommunications infrastructure is lacking.

And so, the race continues to build bigger and better networks.

Network investment roulette

There is, however, a down side to all this network investment. While the investment money makes the job of the telecommunications planning engineer an exciting and interesting one, it carries with it an ominous risk. There are so many different technologies and approaches available, and so few clear, safe infrastructure directions one can take, that the network planner ends up taking a sizable risk. No one can be really sure which technology and which network infrastructure will be the one that survives these times of change and churn. Every time a telco invests in one technological approach, it is in effect betting that its choice will be viable for some time to come.

Let's consider this problem in a little more detail. The original telephone lines, installed by telcos in the late 1800s, consisted of bare copper wires, strung between poles from one location to the next. Each phone required two wires, and had a built-in battery to supply the power necessary to carry the signal. As the number of wires strung between poles became unmanageable, this form of infrastructure became outmoded.

Later, a new technique called *twisted-pair* cabling was developed. Each wire was insulated in plastic sheathing and the two lines necessary to carry one conversation were twisted around each other. This technique allowed many wires to be run through the same cable at the same time and became the industry standard for many years. Many regional carriers still count on these cables to support the bulk of their business.

As technology advanced, the viability of twisted pair also started to wane. Fiber-optic cabling, compressed signaling, multiplexing, and a score of other techniques have made it possible for telcos to carry hundreds and thousands of times as many messages across the same piece of cabling. Of course, each of these techniques requires a different kind of cabling infrastructure and very few of them can make use of the other's efficiencies. In the past, the telco engineers had only one choice regarding cabling (either install the cable or not). Now they have to decide between many different, mutually exclusive options.

Considering the costs involved in recabling major metropolitan areas, how big of a risk are we talking about here? Let's say a company decides to pull out the old twisted-pair cabling and replace it with, say, fiber-optic. The company begins the very long, very expensive process of installing the fiber-optic backbones. Now let's say that, in the meantime, a newer, less expensive technology is invented, one that can carry 10 times more traffic than the fiber-optic and at half the cost. What does the network planner do now?

1. Continue installing the fiber-optic, knowing full well that it may need to be replaced again in the near future?

2. Stop installing fiber-optic and redesign the network again according to the new technology?

3. Some variation or combination of the above?

Management of the network is undoubtedly the most critical component of any telco's strategy.

Marketing strategy development

The development of any telco's corporate strategy, therefore, will include a number of strategies designed to leverage the company's strengths in the three major asset areas (organization, customer, and network) and minimize any weaknesses in those areas. At the same time, the strategies are designed to help position the company in both the short and long term across many different product lines and market segments.

Marketing versus corporate strategy

Looking at the role of the marketing strategy in this light, it should be clear that there is really no easy way to segregate the development of an effective, sound overall corporate strategy from the development of an effective, sound marketing strategy.

It is certainly possible for the marketing department to develop strategic initiatives on their own based on what they believe people will buy with no concern one way or another for the larger network infrastructure and investment issues. However, if those efforts are not timed and coordinated with the organization's plans for the development and capacity of its networks, then it will most likely be a less than successful undertaking. Let's consider a few examples.

Marketing out of synch with organizational capabilities

One of the most popular ways to waste large marketing budgets on meaningless gestures happens when the marketing department musters a huge campaign that results in attraction of a lot more customers than the company's infrastructure is ready to handle. A company with computer information systems geared to manage 100,000 customers, for example, will be in serious trouble if a marketing campaign abruptly increases that number of customers to 150,000. To be successful, any marketing activity that targets the acquisition of large numbers of new customers must be coordinated with the managers of the customer service, provisioning, sales, and network management departments.

Many of the companies who experienced the explosion in cellular telephone services in the past decade saw this kind of unprecedented and unmanageable growth. Many a customer was won and lost again by providers who overestimated their organization's ability to handle the influx of business that expensive marketing campaigns generated.

Marketing out of synch with network planning

Even more abysmal and financially painful problems occur when the marketing department fails to consider the short- and long-term capacity and planning horizons of the network development people. It is important for the marketing strategist to bear in mind that the real product and the real capacity the firm has to offer customers are based on the network as it exists today, and as it is being unfolded into the future.

Marketing can help achieve objectives

The individual corporate strategy that a telecommunications firm puts together—and the role that marketing plays in that strategy—depends on a number of things. Marketing can be recruited to many roles. For example:

- In a conservative and network-dominated environment, marketing will take a clear back seat to the directions set by network planners. The role of marketing in this case is to help the organization maximize the revenue and profit that can be generated by the available network and organizational capacity.

- In the more volatile and opportunistic marketplaces, marketing can become an essential component in the development of an overall strategy. In these situations, the information about market conditions and customer reactions to different technologies allow the marketing strategist to provide key input into the network and organizational investment decisions.

- In economically depressed markets or in markets where technology is obsolescing, marketing can play a key role in minimizing the rate of customer attrition, maximizing the opportunities to migrate existing customers to new products, and easing the cash flow drains that such attrition can create.

Campaigns and telecommunications

The corporate strategy defined by the telecommunications company, as we have described it, can take shape in several different ways.

In some situations, the strategic analysis will dictate that the company's present market profile is fine, which means that upper management feels that the company's image, share of the market, pricing structure, and other critical components are basically sound. In this case, the strategy will be communicated to the marketing department through conservative and traditional means. Budgets will be set, organizational structures adjusted, and goals and objectives related to the operational areas of the business.

In other situations, however, upper management may decide that the company has a serious shortfall in its market presence and that some larger scale, more focused activities are in order. In that case, the corporate strategy could very well become the impetus for the creation of a new corporate large-scale campaign. These corporate campaigns are those major initiatives instigated by upper management to fundamentally shake up the industry. These efforts are designed to force a major realignment of the marketplace's perception of the company and to effect a major shift in the market share, revenue, profit, or wallet share of the firm.

In the development of successful campaigns of this nature, it is critical for the strategist to consider the competitor's weaknesses and the corporation's comparative strengths. An approach that takes advantage of these disparities in each of the three major asset areas (organizational, marketing, and network) will be the most effective. Execution of a campaign of this type often requires that nearly every part of the business change how it operates and refocus on the new objectives.

Campaigns of this type commonly involve the creation of a large number of discrete, small, limited-scope campaigns, which are combined in a focused and coordinated way to accomplish the big campaign's objectives. Two of the largest and most successful campaigns of this type in the U.S. market can help illustrate this point.

MCI's "Friends and Family" campaign

Several years ago, the long-distance market in the United States was a monopoly owned by AT&T. While gradual inroads into that market were being made by Sprint and MCI, AT&T for the most part held on to the market in terms of numbers of customers and in terms of the mind-set of the typical U.S. consumer: There was AT&T and then there were "the others."

The people at MCI, however, had an interesting idea. They looked at AT&T and themselves and compared strengths and weaknesses:

- *MCI:* smaller, newer, more flexible, not much recognition in the market;

- *AT&T:* larger, older, less flexible, a set image in the marketplace (big, corporate).

MCI then looked for a way to mount an offensive that would significantly change its position in the marketplace by playing those strengths and weaknesses.

A revolutionary segmentation scheme

What probably happened is that someone at MCI analyzed how the long-distance market was segmented. At that time, telcos basically viewed the long-distance business as having three segments: large corporate, small business, and consumer industries. All three of these segments were well understood. The corporate and business segments had customers with a high volume of business who were interested in low rates. The consumer segment, however, involved individuals who could not generate the volume needed to warrant cost reductions. It obviously occurred to someone that the consumer market could be looked at from a radically new perspective. Consumers, overall, make many phone calls. If you could somehow get groups of them to work together and negotiate as a block, you could treat them like business customers and offer reduced rates.

The solution MCI came up with was extremely clever and effective. They mounted the "Friends and Family" advertising campaign that recruited people to join the program and then to actually build negotiating blocks around them. MCI directly called thousands of people on the phone and offered greatly reduced long-distance rates if they would simply provide the MCI direct marketer with the names of five other people who could also be offered membership in the group. So, each person contacted provided the MCI telemarketer with the names of five more prospects, their phone numbers, and a personal testimonial from a good friend or family member recommending that they join the program too.

According to the "Friends and Family" approach, large blocks of consumers could be convinced to work together, negotiate as a block, and choose MCI as their long-distance carrier. As a result, large groups of consumers left AT&T at the same time.

Coordination of media and messages

The MCI "Friends and Family" campaign is a good example of a huge success. A sizable advertising budget dedicated to the distribution of radio, television, and newspaper advertising was put into place as the first part of the plan. These advertisements were used to recruit the initial contact people and to generate a general awareness of the program in the public's mind. Then the segmentation recruiting process began, with the lion's share of the effort carried out by the telemarketing organization. Ultimately, the campaign led to some significant shifts in long-distance business allocations in the United States.

Organizational alignment—key to success

Of course, to be effective, the strategists behind the "Friends and Family" campaign had to consider the entire organization and be sure that the infrastructure was in place to support it. Remember, the consumer market was considered to be one large block of customers with little or no price-negotiating capabilities. Also recall that the "Friends and Family" program stipulated that the reduced rates only applied to the people enrolled in the program. Consequently, to be able to support the promised discounts, MCI had to do some serious investigation and reengineering of their billing systems. Here again, the strategists at MCI were able to take advantage of one of their strengths (smaller size, newer and more flexible systems) and AT&T's weaknesses (large, cumbersome, contumacious information systems) to create an advantage that AT&T still cannot match.

AT&T Digital One Rate

Sophisticated segmentation schemes and complicated marketing programs are only one form of strategic campaign initiative, however. Other companies with different parameters will develop altogether different kinds of campaigns. A good example of this is AT&T's Digital One Rate campaign. This campaign, run by AT&T Wireless Services, sets the stage for a significant shift in market share through the strategic leveraging of its network assets.

While AT&T may hold the majority of the business in the long-distance marketplace, the wireless services organization has been less than dominant. It is clear, however, that someone at AT&T Wireless Services realized that they owned a significant strategic advantage that could be leveraged if they could change the way the industry and consumers looked at wireless phone service.

Consumer perceptions of wireless in the United States

In the U.S. wireless industry, the consumer's perception of costs and services is dominated by not one, but three separate charges. A customer is expected to pay:

- A per minute charge for connect time;

- Any long-distance charges (accrued through a long-distance carrier);

- Any roaming charges (applied by a wireless carrier who "hosts" the wireless phone when the caller is outside of the coverage area of the primary provider).

From the consumer's perspective, two of these charges, the wireless connect time and the long distance, are controllable and negotiable expenses. A consumer can shop for a cellular provider and a long-distance provider and get the best rates possible. When it comes to roaming charges, however, the U.S. consumer is out of luck. Those costs are imposed, not negotiated, and because the consumer can do very little about it, they tend to be excessive in most cases. From the consumer's point of view, therefore, roaming charges represent a large, uncontrollable, imposed expense that they will go out of their way to avoid.

Leveraging network strengths

When the strategists at AT&T Wireless Services compared strategic assets (organization, marketing, and network) in the wireless market, they saw that one of the three pricing components that customers are concerned about happens to be one where AT&T has a strategic advantage. Long distance is one of AT&T's strength, while the other wireless companies for the most part (at least at the time the program was launched) had no such capability. They must route the long-distance traffic to another provider. AT&T Wireless Services realized it was in a position to offer a bundled package of services that includes local wireless, long distance, and roaming at a lower cost than the competitors.

Flat-rate wireless

The result was the AT&T Digital One Rate program. Under this plan, consumers are able to get a set, low-cost cellular hookup without added roaming and long-distance charges—one rate for all three service components.

The short-term effect of this program has been the phenomenal growth in subscribership for AT&T Wireless Services. The longer term result will undoubtedly be more long-distance/wireless partnerships and mergers and eventually a shift in the entire wireless industry to a more flat-rate kind of service.

5

Media, Messages, and Outsourcing

The medium is the message.
Marshall McLuhan

To understand the overall marketing process and its component parts we now look at two of the other key ingredients of the marketing process, the media and the messages.

Media

Our discussion of the marketing process has thus far focused on whom to market to, but we have said very little about how to address our target population. Marketers actually have a dizzying array of options to choose from when they set out to influence the behavior of customers and prospects. We use the term *media* to describe the various vehicles that marketers use to send a message or create an interaction with the customer or prospect. Traditionally, the world of marketing defines the following major categories and subcategories of media.

Advertising (broadcast) media

The most obvious and most easily recognizable form of media utilized by the telecommunications marketer is advertising or broadcast media. Advertising, as we all know, is communication of a prepared message that is broadcast to a large audience. The most common categories of advertising media are discussed in the following subsections.

Television and radio

When people think about marketing and advertising, the first image that comes to mind is that of the powerful messages broadcast daily via television and radio. These forms of advertising give a company instant, powerful access to millions of potential and current customers in a way that is often very difficult to ignore.

Print

The second most recognizable form of advertising comes through print media such as newspapers and magazines. When marketers want to get more specialized or local in their approach, or when their message needs to be more personalized, they often resort to print media to broadcast their message.

Display

Another form of broadcast media that allows for even more geographically specific targeting is display media. This kind of advertising can be executed as billboards, place-ads in trains and buses or on grocery store shopping carts, or ads painted on the sides of buildings.

Key characteristics of broadcast media

While broadcast media are an extremely popular form of marketing outreach, they are not always the best approach. They have a number of shortcomings.

1. *Inefficiency.* A major drawback of broadcast media is that the message is likely to be delivered to many more people that you intend to reach. You typically pay to advertise to a much broader audience than the one in which you are interested.

2. *Cost.* Broadcast media are very expensive and advertising budgets can quickly move into the millions of dollars.

3. *Inability to tune message.* With broadcast media there is no way to customize the message. Any offer you make to the public in this manner must be for all people. This means that you must be prepared to deal with everyone who responds whether it is the kind of customer you want to attract or not.

4. *Inability to control impact.* With broadcast media there are no second chances. If your message is interpreted in the wrong way, then you have broadcast negative advertising to a large group of people with very little chance of taking the message back.

5. *Inability to test responses.* It is impossible to know exactly how much impact any message is having. All you can do is look at the passive results and assume that the advertising is contributing somehow, which is fine until you get involved in the delivery of many different advertising messages through many different media. At that point, you cannot determine which activity is contributing the most to the results.

Promotion

In contrast to the impersonal and far-reaching broadcast media is the very personalized public relations and community outreach known collectively as promotional activities. This concept includes all of those activities through which telecommunications firms try to participate in the activities of the communities to which they cater (both business and consumer communities).

For example, the small and medium business division of a large telco might acquire booth space at local trade shows where it demonstrates to business people how their operations can be improved through extended business services. At the same time, the consumer division might sponsor events at the local county fair to show consumers the different kinds of personalized services they have to offer. Sponsoring community and sporting events and notifying local new agencies of community-related services and contributions are all promotions that help accomplish marketing objectives.

Direct marketing

Without a doubt, however, the single most popular and efficient means of marketing in the modern world of telecommunications is direct marketing. *Direct marketing* is the term used to describe the process of sending specific messages to specific customers through mail, telephone, fax, or e-mail.

History of direct marketing

The direct marketing industry itself, of course, is not new. Marketing by mail in particular has been a retail institution in the United States since the colonial days. As time went on and as society progressed, so too did direct marketing. The golden age of direct marketing occurred in the late 1800s and early 1900s when retailers such as Sears, Roebuck, and Montgomery Ward revolutionized the world of commerce and fueled the great Western expansion through their retail catalog industry.

Direct marketing today

Since the beginning of the twentieth century, the industry has not been idle. Direct marketing through catalog houses, direct mail organizations, and the ever popular phone solicitation industry has continued to grow and expand in both economic clout (representing a multibillion dollar a year industry) and in the sophistication of techniques employed.

One-to-one marketing

The direct marketing industry and its discipline have been enjoying a renaissance of popularity in the eyes of most businesses. The reasons for this newfound respect for the tried and true disciplines are relatively simple and straightforward. While other businesses have spent the past 100 years developing disciplines and expertise in the broadcast marketing activities, the direct marketers have been concentrating on keeping better track of individual customers and have focused their marketing efforts on one customer at a time. They have been learning how to manage large populations of individuals and how to optimize the service and revenue that can be generated for and from each.

The new age of marketing

As computer technology continued to grow in power and capacity, something happened in the business world. Suddenly, companies found that modern databases and data warehouses made tracking of individual customers possible. Computer systems now made it feasible to establish personal relationships with millions of people and keep track of them. Consequently, more and more companies are turning to the direct marketers and their approaches to learn how to run their own businesses on this same one-to-one customer basis.

Role of direct marketing in telecommunications

The telecommunications industry has by no means been immune to the attractiveness that the one-to-one direct marketing approach offers. There are several reasons for this:

1. Telecommunications companies already have one-to-one personal relationships with their customers by default. No firm has more information about the personal habits, movements, and activities of its customers than a phone company. After all, every single phone call you make, where to or from and for how long, is a matter of record. Because the personal relationship and information are already in place, the telecommunications firms have a huge head start in the mastery and management of individual customer populations.

2. Telecommunications firms have recently gone through a period of extremely volatile competitive positioning. Companies that never had to compete before needed to learn, very quickly, exactly who their customers are and what they want. Telecommunications companies have a very serious need to understand customer desires and behaviors if they are to be competitive in the future.

3. In countries that have already undergone deregulation and competitive pressure, telcos have learned that direct marketing disciplines are powerful foundations on which to build successful strategies.

4. Direct marketing allows the marketer to function with pinpoint accuracy. Broadcast and promotional media reach many people at a low cost per person, but this approach provides for almost no precision. A broadcast is a message to everyone. A direct marketing message addresses an individual.

5. Direct marketing builds accountability, feedback, and marketing profitability into the system. With the broadcast and promotional media, the marketer can never truly track how effective the campaigns are or determine where improvement lies. The results of broadcast and promotional activities are by definition interpretive and guesswork. Direct marketing, on the other hand, allows for the specific management of campaigns with almost immediately measurable results.

For these reasons and more, we are seeing a tremendous shift of telecommunications marketing efforts from the broadcast and promotional areas to direct marketing.

Organization as medium

The tools available to the marketer for crafting and delivering a specific message to a specific market segment are not limited to the external, traditional advertising media. In fact, one of the most powerful tools is the telecommunications organization itself. Any part of the telco that has direct contact with customers, such as customer service or the sales force, can be used in organizing a campaign. In fact, crafting a specific marketing approach might involve nothing more than a simple change in credit policy, in responsiveness to consumer complaints, or in the creation of a new department to focus on the needs of small home-based businesses.

Hybrid activities

Thus far, we have talked about the different media employed by the marketer as if they were mutually exclusive, that the choice is either advertising or promotion or direct marketing. In reality, however, marketers usually combine some or all of these media when developing a specific campaign.

Ads in bills

One of the most cost-effective forms of hybridized marketing activity is to include advertisements with the billing materials that are sent to customers. This technique effectively combines direct marketing with customer service at a very low additional cost to the firm.

Customer service solicitation

Another popular technique is to provide customer service representatives with scripted sales approaches. When customers call regarding a particular kind of problem, they are offered alternative solutions that involve upgrading or changing of the customers' current product set.

Advertising to generate sales force leads

Another extremely effective combined technique is to launch an advertising campaign (either broadcast or direct marketing) to generate qualified leads for the sales force to follow up on.

Massive campaigns

The large-scale, industry-shaking campaigns that are promoted into the market usually integrate and coordinate messages and activities using all of the available media.

The effect of media selection on the marketing process

Selection of the media that the telecommunications firm will use is an important part of the job of the people taking part in the marketing process. The actual point in the process where media selection occurs will vary depending on the situation. Sometimes, the sponsor of a specific marketing activity will have a media in mind even before forming the project team. In this case, media selection occurs during the prioritization and goal-setting phase and becomes one of the constraints dealt with by the team. Other times, the people on the project team will decide during the modeling process which media will be the most effective. In many cases, the specific media will not be selected until actual campaign development is under way.

The point within the process where media decisions are made can be critical to the efficiency of the overall marketing process however. At times, certain media require that marketers develop specific kinds of models in order to validate their assumptions. The model for a direct marketing campaign will have different segmentation, optimization, and feedback measurement criteria than a broadcast campaign.

The differences become even more pronounced as we move into campaign development. For a broadcast campaign, the marketers generally work with outside agencies to create advertisements, conduct focus groups, and prepare other kinds of premarketing and postmarketing measures and checkpoints. For a direct marketing campaign, on the other hand, the marketers will focus on customer lists, attribute management, and execution of the direct marketing disciplines.

Messages and missions: the relationship between marketing, sales, and customer service

In the telecommunications industry, organizational messages and missions tend to overlap, which is one of the reasons the marketing process can become so confusing. Although the differences between marketing, sales, and customer service can be clearly defined, the lines between them can blur when faced with day-to-day reality. In general, we could say that responsibilities of the groups are as follows:

- *Marketing:* market research and advertising (mass, direct, and promotion);

- *Sales:* phone and personal contact with customers and order taking;

- *Customer service:* answering customer questions and resolving customer difficulties.

Of course, it gets much more complicated than that in the real world.

The role of channels

In the old days of telecommunications, marketing did very little and sales was about answering the phone to take orders. Customer service was where most of the emphasis was placed because public opinion and regulatory scrutiny were the most important issues.

Today's unregulated telco, however, is one where marketing plays a pivotal role and sales involves not only phone sales and local office personnel, but in many cases includes complete alternative sales channels. Jobbers, retail outlets, direct marketing organizations, and other channels create a cavalcade of conflicting, competing, and colliding sales efforts, all desperately trying to get the customer's attention and the next order. Telecommunications has changed from a single sales channel industry into a multichannel industry almost overnight.

Channel conflict and resolution

Organizations make use of so many different channels for good reasons. More channels mean more ways for the customer to get the message and buy the product. Of course, more channels can also mean more confusion, especially in the area of marketing messages.

The Costa model of channel objectives

Originally developed in the 1980s by Dr. Paulo Costa, the *Costa model of channel objectives* attempts to identify and isolate the different missions that various channels pursue (Figure 5.1). It tries to help the marketer avoid potential conflicts by associating the appropriate mission with the channel best suited to it.

The Costa model describes the different missions that a marketing/sales effort has as follows:

1. *Awareness.* The earliest form of contact between your company and a prospect is awareness. After all, if prospects do not know that the company exists, they cannot buy with confidence. You must establish awareness before you can accomplish anything else.

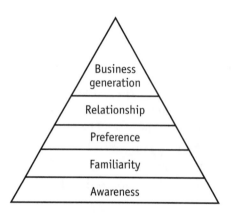

Figure 5.1 The Costa model of channel objectives.

2. *Familiarity*. After the prospects are aware of your organization, the next step is to get them to understand who you are and what you are offering. This is the point in the relationship when information, both factual and emotional, is communicated to the prospects in order to establish an association between an image of your company and meeting specific needs or wants.

3. *Preference*. Once the prospects know who you are and what you are selling, next you want them to prefer you to your competitors. Preference is accomplished through competitive pricing, service levels, and many other means.

4. *Relationship*. Once prospects acquire a preference for your company and product or service, they begin to develop a relationship with you. Give and take occurs, a comfort level exists, and a willingness to share information is attained.

5. *Generation of business*. Finally, and only after the other levels have been reached, will the prospect sign on the dotted line and become an actual customer.

Single sourcing of messages

Let's consider a few examples to see how the Costa model helps manage the messages that different channels deliver.

Sales force as sole agent

It is possible, in fact common in some industries, for the company to do little or no advertising of any kind. These firms depend solely on the work of direct salespeople to make their business work. Companies such as Avon, Shaklee, Fuller Brush, and Encyclopedia Britannica use their direct sales force to accomplish all of the objectives of this model.

Retail channel as sole agent

A wireless phone company may use local retailers to sell their products and services. In this case, the telco leverages the existing awareness, familiarity, preference, and relationship to persuade prospects to make their final buying decision.

Direct marketing as sole agent

In other cases, companies have relied solely on direct mail or outbound phone solicitation to generate the bulk of their business.

"Best practices" channel utilization

Most telcos today try to use as many channels as possible to get their message out, and the marketers must understand the relationship of each channel to the customer and determine which channel to use to deliver what part of the message. In general, telecommunications organizations find that the best fit for each message tends to be as discussed next for each channel.

Awareness

By far the easiest and most cost-effective way to create awareness is through the mass media. If the telco's target is a large percentage of the available market, then mass marketing will be an essential ingredient of the strategy.

Familiarity

Conveying familiarity is a little more difficult. Selecting a channel for establishing an understanding on the customer's part depends on the complexity. For example, companies with very complicated rate structures often create confusion among potential customers when they try to communicate those structures via mass media. For the most part, the simpler the nature of the understanding, the better mass media is for delivering it. The more complicated the message, the more individualized and personal the delivery should be.

Preference

Establishment of preference is usually a function of the message. Price preference, service preference, and other logical preferences can be delivered with equal effectiveness through all channels. The right-brain, subconscious preference messages, however, such as status, prestige, and exclusivity, depend very much on the delivery channel. For example, a "wealthy" message delivered via a television ad will turn off a customer looking for discount rates.

Relationship

To establish the buying relationship, some form of personal contact, either via mail, phone, or in person, is usually required. The relationship contact point is the ultimate portal to the sale.

Business generation

Finally, at the point of business generation when the customer actually places an order, it is essential that some personal form of contact be made.

Messages: what do you want to convey?

Although use of the right media is essential if you are to communicate successfully with prospects, it is useless if you do not have a powerful message. The whole art and mystery of creating marketing messages is a subject best left to other books and other authors to explore fully. Suffice it to say that the creation of the message is certainly as important as any of the other aspects of the process.

Direct versus indirect messages

An interesting aspect of marketing message development is that it can be constructed to send either a direct or an indirect message. Don't assume that the intention of all marketing activities is to produce an immediate action. Sometimes, messages will indeed be explicit in what the marketer wants the prospect to do. Other times, the message will simply imply, lead, or assume what kinds of actions will occur as a result of the stimuli presented.

Action versus attitude messages

One way to distinguish between the different kinds of messages is to consider whether the intention of the message is to inspire the prospect to *do* something or to *feel* something. In other words, is the intent of the message to change the actions of the targeted audience or their attitudes?

Tying the messages to the objectives

Telecommunications marketing teams are usually concerned with only a few key messages, those that relate directly to the customer behaviors or attitudes that the marketing project sponsor has singled out for the project. We therefore expect to see messages that attempt to effect the following:

1. *Acquisition:* messages that entice people to subscribe to various services;

2. *Retention:* messages that convince customers not to leave the service they already have;

3. *Wallet share/utilization:* messages that encourage customers to increase the use of the services they have.

Reasons for subscribing

The message "*Subscribe because ...*" is probably the most common form of telecommunications marketing, especially in the long-distance and wireless industries. These directives are primarily concerned with getting customers to either start using a new service or switch from an existing one. People can be encouraged to switch to an alternative for many reasons including (1) lower costs, (2) better service, (3) faster connections, and (4) better quality.

Reasons for using

Utilization advertising, marketing campaigns that encourage people to use more of a particular service, has not been greatly evidenced in today's competitive telecommunications market. It was pretty much the only kind of advertising a telco engaged in when they didn't have to compete for customers. Utilization advertising helped make customers aware of who their telecommunications provider was and developed a loyalty to that provider, which came in handy when competition was made possible.

As the markets mature and as the churn and market share battles become less important, utilization advertising will most likely become an important tool again for the telco.

Reasons for loyalty

"Don't leave" is without a doubt the most important message that telcos are interested in sending to customers. Churn, the turnover of customers to other carriers, is the single biggest problem faced by most wireless and many long-distance carriers today. Accordingly, the development of "don't change" messages is of critical significance.

Role of segmentation in message development

To craft the best messages, the marketer needs to understand the important details of the customers to be approached. Who are these people? What do they like? What is important to them? What motivates them? All of this information is vital, if the message is going to be effective.

It is here, in the area of message formulation, that segmentation plays a critical role. Armed with good, accurate segmentation information, the marketer can prepare messages that will attract and persuade people with accuracy.

Messages delivered at many levels

Effectively communicating the right message to the right people at the right time is a challenging job. Marketing messages, unfortunately, are subject to interpretation by the audience and that audience can be extremely unpredictable in how it responds.

Multicultural interpretation challenges

Perhaps the easiest way to highlight the problems that message interpretation creates is to think about how different cultures may respond to the same message. The marketing industry is full of stories of a message that worked well in one cultural setting and was applied to another cultural setting with disastrous results. The use of some colors or product names can be inappropriate, humorous, or even have calamitous consequences in cross-cultural situations. The color black, for example, may be interpreted as neutral, positive, or negative depending on the culture. Clever product names such as *Burger King* may be quite appropriate in, say, the United States, but may come across as demeaning or insulting in cultures with reigning monarchs.

The key for the marketer, then, is to be sure that messages will be interpreted as expected.

Conscious and subconscious interpretation

Sending effective messages is made especially difficult because people will respond to them on many levels. A person's conscious response of "No, I am not interested" might hide the subconscious response "I feel insulted by this ad." People are rarely aware of their secondary responses to messages.

Examples of events and interpretations

Let's look at some examples of marketing activities and see how one event delivers multiple messages.

Participation in local events

A telco's participation in a small local business fair can be used to send many simultaneous messages. The explicit message for their participation will be prominent. For example, the booth can be set up to entice more people to sign up for business-related services. The message of "Subscribe now because … " will be clear.

At the same time, the very fact that the telco has bothered to show up at the event sends another, subtler message of "We care about you. Your continued support is important to us." In other words, it also sends a strong message of "Don't change because … we support you and your business activities."

National television advertising campaigns

Glamorous national advertising campaigns also send more than one message at a time. The explicit aim is again a "Subscribe because … " message and will be clear to the targeted audience.

At the same time, a large campaign like this communicates a world of information to customers about why staying with the current service is the right thing to do. This kind of message is reinforcement advertising and assures customers that their earlier purchase decision was a good idea.

Conflicted messages (a.k.a. antimarketing)

Sending multiple messages can also have a serious down side such as when customers get conflicting messages. This often happens when a telco uses multiple media sponsored by different parts of the organization. Sending too many messages can confuse customers about who the company is and what it is offering.

The role of the marketing database in message management

To manage the situations when multiple messages are bombarding the marketplace, many organizations are implementing campaign management software. Campaign management is one of the major components of a marketing database environment, the role of which is to keep track of the different messages that are sent to different people at different times. It tracks contacts and measures the effectiveness of the campaigns.

The role of testing in message development

One aspect of message management is keeping track of who gets what message and making sure that conflicting messages are minimized. The other equally important one addresses the development of specific models that substantiate that the intended message is in fact what the prospect or customer receives.

Message testing metrics and models

Because of the importance and complexity of the messages sent via different media, marketers will often test them on a limited audience to assess their impact before deploying them on a very large scale (when the effects of a poorly constructed message can be disastrous). These tests can be very small and casual or large and formal.

For broadcast media such as television or radio it is common to run consumer focus groups or test fly ads and then interview the test consumers on what they think of the message and how it made them feel. These kinds of activities are provided to marketers through specialized testing firms and are usually given to relatively small representative samples of people.

Another form of test is often used as part of the direct marketing approach. Direct marketers frequently run test campaigns in which a small, representative sample of the entire population is approached with a telephone or mail solicitation. The actual sales results are tracked very closely. This testing allows the marketer to estimate how effective the program will be and to test the results of different messages before committing to a large-scale, major marketing program.

Media and message development: the role of agencies and outsourcing

The job of selecting media and preparing the messages to be sent through those media is an extremely complicated, sophisticated, and specialized job. Because of this, telcos find that they need to outsource many pieces of the marketing message development puzzle. Being able to call on marketing specialists allows the telco to get the best possible quality of marketing media and message at a reasonable cost. Cohesiveness, however, tends to suffer because of the many different outsourcers that need to be included. Assistance by these external sources can create much confusion and inefficiency in the execution of many telco marketing processes. The following subsections describes a few of the more often employed external experts and the roles they usually serve.

Advertising agencies

Advertising agencies are typically associated most closely with broadcast media. They usually provide the following marketing support services:

1. Develop broadcast campaigns.

2. Perform segmentation and other modeling activities.

3. Create messages and select media.

4. Test messages before broadcast.

5. Book advertisements and negotiate rates.

Advertising agencies will often take on much of the responsibility for large campaigns. In fact, in some cases upper management might give the whole campaign to the agency and leave the marketing department completely out it. At other times, the agency is used to support only specific parts of the process.

Database marketing firms

The one-stop-shopping typically offered by advertising agencies for the broadcast media area is applied to the running of marketing campaigns by database marketing firms. Full-service database marketing firms can handle these tasks:

1. Develop direct marketing campaigns.

2. Produce execution models and segmentation studies.

3. Create the messages and media (print or telephone script).

4. Identify and construct mailing lists.

5. Run test campaigns for effectiveness assessment.

6. Print the actual materials.

7. Mail the materials.

8. Execute the phone campaign using their own phone solicitation staff.

9. Respond to the leads generated.

10. Deliver products and services to the customers.

While the full-service database marketing firm tries to do everything for the telco, their services are usually employed on a piecemeal basis with the telco marketing team choosing to use them to supplement their activities as deemed appropriate.

Specialist consultants and services

In addition to the two major types of marketing service providers (agencies and database marketing firms), there are literally thousands of specialized consulting and services firms, each working to provide different parts of the services required to put successful marketing campaigns together. You can hire consultants who can develop models, create messages, test campaigns on consumers, and handle every other part of the process. These specialists work either directly for the telco or indirectly through the agencies and database marketing firms.

6

Direct Marketing

The shortest distance between two points is a straight line.
Euclid

One customer at a time.
Don Peppers

In the last chapter, we introduced the concept of marketing media and discussed some of the more important media that telecommunications firms use. We also talked about the role of direct marketing, that is, the use of phone and mail solicitation as a core medium that is used increasingly by telecommunications firms. In this chapter, we will take a much closer look at direct marketing and how it can be used to support telecommunications marketing activities.

Definition of terms

The first problem encountered when discussing the direct marketing/database marketing arena is the problem created by the confusion in terms. Many people define a multitude of concepts using the same terminology. As a result, it becomes difficult to sort it all out. Let us begin with a restatement of what the core definitions actually are.

Direct marketing

The term *direct marketing* is used to describe a form of marketing that delivers its messages to prospects and customers via a direct, personal approach. In other words, direct marketing is marketing via mail or telephone. The only other way to reach people is via face-to-face communication and that is usually called *sales*.

In direct marketing the two core ingredients you need are (1) a list of people's names and locators (either addresses or phone numbers) and (2) a service or a product to sell them.

The direct marketing industry

While the process of direct marketing can be done by anyone with an address book and some postage stamps, the direct marketing industry specializes in this type of marketing. It is an integral part of the much larger retail industry.

In general, all companies that make their living by preparing and mailing catalogs or by conducting outbound phone solicitation are part of this extremely profitable industry niche.

Database marketing

Although companies in the direct marketing industry have exercised the direct marketing process (i.e., the use of name lists to sell to people) for many years, it is only recently that they have started to make use of databases.

Direct marketing, as practiced in the "old school," uses lists of names for all the work required. In today's modern technology world, however, more companies are beginning to replace their old-fashioned lists of customer names with databases full of customer information. Database marketing, then, is nothing more than running a direct marketing process with databases instead of lists.

Database management in non-direct-marketing firms

The direct marketing industry is in the process of modernizing its approaches to make use of newer database technologies. The business world in general, and telecommunications firms specifically, are finding that the construction of customer name-based marketing databases are similar in construction and purpose to the ones built by direct marketing firms. These can be critical to their ability to execute effective marketing campaigns.

We find that more telecommunications companies are investing in the construction of these new marketing database systems. These systems tend to combine the needs of the marketing organization through the execution of the marketing process (prioritization and goal setting, modeling, campaign development, campaign execution, analysis and feedback). This provides the marketer with a powerful and dynamic means to track the behavior of individual customers and the marketing messages they receive.

Companies who build and use marketing database systems of this type usually find that the initial investment is paid back many times over in marketing effectiveness.

Customer relationship marketing or customer relationship management

In recent years, a new type of system, known as a customer relationship management or marketing system, has also become popular. These systems contain the core functionality provided by marketing databases, and extend their functionality beyond the marketing process, to sales, customer service, and other parts of the business.

With a customer relationship management system, everyone in the organization can track and report on all contacts they have with the customer. Armed with this comprehensive information, the business is better able to manage how the customer is treated and evaluate how to more effectively meet that customer's needs (see Figure 6.1).

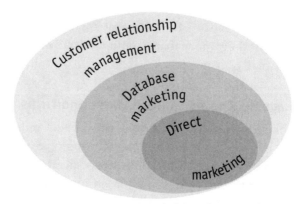

Figure 6.1 Relationship of direct marketing to other approaches.

You cannot have a CRM system without a marketing database, although the CRM system includes other items. At the same time, you cannot have a marketing database without making use of the principles and fundamentals of direct marketing. Again, the marketing database does more than simply manage the lists with which a direct marketing system is concerned.

Direct marketing in consumer versus commercial business

While many people believe that direct marketing is mostly a consumer-based industry, nothing is further from the truth. A very large and lucrative branch of the direct marketing industry, known as the "business-to-business" or "B2B" industry, makes use of the same disciplines to support the execution of direct marketing to commercial customers.

Many examples in this book refer to the consumer part of the business, but this does not mean that the same marketing, marketing database principles, and approaches do not apply equally to the business front. The only difference between consumer database marketing and business database marketing is the way in which customers are grouped. In the consumer database, customers are associated with buying units called *households*. In business database marketing, customers are associated with buying units called *businesses*. Everything else is the same.

Direct marketing and the telecommunications firm

Telecommunications firms around the world are finding that the direct marketing approaches we are discussing here, both in the pursuit of consumer and commercial marketplaces, hold potential for increased efficiency and market share penetration. In the following sections, we will explore how these direct marketing disciplines work.

The direct marketing process

Given our understanding of direct marketing and the hierarchy of systems (direct marketing, database marketing, and customer relationship management) that it supports, we are ready to develop a basic understanding of how the process works.

Simple direct marketing

We will start by examining the core functionality that makes up the direct marketing process. Then we will look at some of the interesting challenges and opportunities that the process creates, and see how direct and database marketers have addressed them.

The basic process

For direct marketing, all you need is a list and a product. The basic process is something like the following:

1. *Get a list.* First, the marketer needs a list of the names of the people that are likely buyers. This list can be extracted from the telephone book, purchased from a firm that specializes in creating lists, or developed by the telco in house.

2. *Prepare a message and media combination.* The marketer puts together a message in media form (for example, a full-color advertisement of the item being offered).

3. *Send the message.* The marketer will then "execute," sending the piece to everyone on the list.

4. *Fulfill orders.* The customers receive the ads and send in their orders. The company is now ready to "fulfill" the order (send the product to the customer and collect the money).

Direct marketing campaigns versus marketing campaigns

An interesting vocabulary issue arises when we combine the world of direct marketing with the marketing world at large. In direct marketing, the process of (1) identifying a group of people to mail to, (2) creating materials to support them, (3) sending those materials, and (4) answering customer requests is referred to as *running an individual campaign*. In other words, the execution of one mailing is called a campaign. However, in the marketing world at large, one campaign will typically consist of many individual campaigns. It is therefore critical that we are clear about the context when we discuss campaigns.

The core direct marketing disciplines

Of course, the simplistic direct marketing process that we just described can become more complicated in a hurry. When you start to deal with many products, many people, numbers of lists, and lots of campaigns, you can quickly become overwhelmed by the process.

Over the years, direct marketers have broken the process into a series of distinct, easily managed, and well understood direct marketing disciplines. These include the following.

List (name) management

While our simplistic view of direct marketing calls for us to merely "get a list of prospects and mail a message to them," in reality the direct marketer has hundreds of potential list sources from which to choose. Because lists will have different strengths to offer, marketers often find that they need to work with a number of lists at the same time. The process of identifying, standardizing, and managing multiple lists of names is known as the *list management process.*

List selection

Identifying alternative sources for names is the first challenge faced by the direct marketer. The second challenge is to look at all of the available names and choose the ones on which the campaign should focus. Typically, when direct marketers get ready to execute a campaign, they create a special, hybrid list, combining the best prospects from several list sources. The process used to select a list, or combination of lists, that is to be used as the basis for a campaign can be simple or very complex. This depends on the sophistication of the marketer and the nature of the campaign. The process of deciding on the characteristics for this list, however, ultimately comes down to the marketers making basic segmentation decisions (choosing the people on which to target the campaign), based on their previous behavior or characteristics. (See the chapters about the segmentation process for details.) This composite list of the "best candidates" for a campaign becomes the "base list" for the campaign.

Name selection and scoring

After a list of the most likely candidates for a particular offer have been identified, some direct marketers move directly from list selection to the actual execution of a campaign. In many cases, however, the direct marketer will want a more precise prediction about how well the mailing or phone campaign will run before committing to it. For example, if each message sent to a customer costs $10 to deliver, then they want to be relatively sure that the customers are worth contacting before including them on the final selection list.

When marketers decide to get extremely precise with these predictions, they execute a process known as *scoring*. Scoring is the process of analyzing all names on a list and statistically determining how likely individuals are to buy and/or how much they are likely to spend in response to a particular campaign. A list of prospects who are all considered equally likely to buy is turned into a specifically scored list. This provides the marketer with a ranked listing of those same people, revealing the statistical likelihood of the buying behavior of each one. Through scoring techniques, marketers are able to take a specific list and determine exactly who on that list should be approached, based on their statistical likelihood of spending.

Testing

Many times, if the campaign being considered is large or expensive enough, direct marketers will run small test campaigns to determine how effective the larger scale campaign will be. Through testing, marketers get the chance to vary the message, the medium, and their segmentation assumptions before committing all resources allotted for a given campaign to a particular message/medium combination.

Contact management

In addition to the efforts that marketers expend in trying to get the maximum benefit from individual campaigns, they encounter another set of problems when the same list of names is used to support several campaigns at the same time. Marketers need to worry about two additional problems when many campaigns are run. The first problem is that they need some way to measure the cumulative effect of multiple campaigns on the same person. In other words, if the same person is the recipient of 12 different mailings, and they buy after they receive the 12th one, how do we know which of the 12 mailings, or what combination of mailings, led to the purchase decision? The second problem is that if an individual receives too many mailings (or phone calls) he or she can become angry at the marketing companies. In either of these cases, the marketer needs the ability to track the number of times and ways in which each person was contacted.

Campaign tracking

Not only do marketers need to keep track of the different contacts made to individual customers, they also need to know exactly how effective each campaign is. Unsophisticated organizations will simply execute direct mail campaigns and then just wait to see if sales go up or not. Sophisticated direct marketers keep track of which customers respond to which messages and in what way. This detailed tracking of the effectiveness of campaigns on the individual is one of the most powerful sources of information for the marketer when attempting to build future campaigns.

Campaign management

The overall process of all of these different jobs, focusing on the management of lists of names, and the execution of campaigns is known generically as the *campaign management process* (Figure 6.2).

Fitting direct marketing into the overall marketing process

The direct marketing process we just described is, in fact, a *complete* marketing process for companies who make their living that way (the catalog houses and other types of direct marketing firms).

However, when a company whose primary business is something other than the creation and promotion of catalogs and other typical direct marketing media decides to try to tap into the immense power and flexibility that direct marketing has to offer, problems can result. Serious organizational and conceptual problems are usually the result when companies try to merge the highly disciplined and tightly focused direct marketing model with the much more flexible and multifaceted telecommunications model.

When this merger is handled properly, there is an amazing synergy between the two models. The result when this merger is not handled properly can be a chaotic potpourri of conflicting ideas, objectives, and procedures.

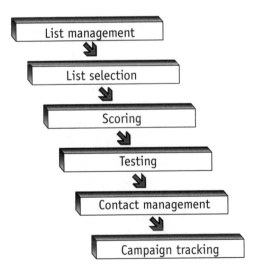

Figure 6.2 The campaign management process.

The telecommunications marketing process against a direct marketing backdrop

If we were to look at the basic telecommunications marketing process, as described in Chapters 2 and 3, and map each of the individual components of the direct marketing process against it, we could see how these two processes could be blended together.

The basic steps in the direct marketing process are list management, list selection, scoring, testing, campaign execution, contact management, and campaign tracking. The basic steps in the telecommunications marketing process include (1) prioritization and goal setting, (2) modeling, (3) campaign development, (4) campaign execution, and (5) ongoing measurement and feedback. Let's consider how these line up with each other.

Prioritization and goal setting

While the marketing organization may at some point decide that direct marketing will be the selected medium to carry the organization's message to the customer, at the early stages of the marketing process that decision will not yet have been made. At this point, direct marketing is only an option to be considered, and the direct marketing process itself will be neither involved nor invoked.

Modeling

When the marketing organization decides on specific objectives to be met, the processes of direct marketing begin to come into play. During the modeling process, marketers will begin to develop specific ideas about the messages they want to deliver and the media they want to use. At some point in this process, they will need to develop specific numbers, proving to management that the media and messages selected will be effective ones.

Any marketing campaign that includes direct marketing media must therefore involve the execution of the first and second parts of the direct marketing process (list management and list selection). Indeed the whole objective of the modeling process, in the direct marketing case, will be an attempt to determine which lists (or subsets of lists) the marketer intends to use in support of the campaign, along with a set of ancillary data to back up the decision.

In conclusion, the output of the modeling process is a decision to send a particular message, using a particular medium, to a specific segment of people, and the mathematical and/or logical proof that indicates how effective this particular combination is expected to be.

Campaign development

After the modeling process has been completed, and the direct marketing medium has been selected, the marketer can become serious about the use of the lists. Through scoring and testing, the third and fourth steps in the direct marketing process, the marketer takes the generalized assumptions made during the modeling step, and turns them into specific, measurable predictions.

It is here that the line between modeling and campaign development often blurs. For example, in some situations, the development of scoring and testing models presents the marketer with new information, which will ultimately cause them to return to the "drawing board" and revisit the modeling process. At other times, the results will confirm what was proposed during the initial investigation's modeling step. Ultimately, however, the output of the campaign development process is a specific scored list of individuals to whom a particular message will be sent.

Campaign execution

The execution of a direct marketing campaign is standard whether it occurs within a telco or within a catalog house. The media and messages are prepared and delivered to customers.

Ongoing measurement and feedback

It is here, in the area of ongoing measurement and feedback, that the direct marketing discipline can be especially useful to the telecommunications marketer. While the broadcast and promotion media leave the marketer with little tangible framework for assessing the effectiveness of a campaign, the fact that the direct marketer made specific offers to specific people or companies means that it will be possible to evaluate just how effective the process has been.

The direct marketing steps of contact management (keeping track of who was given what message) and campaign tracking (keeping track of the results of campaigns in a measured, methodological manner) makes direct marketing especially attractive.

Applying direct marketing to telecommunications marketing

The core processes that define the discipline of direct marketing have proven incredibly useful to telecommunications marketers for multiple reasons. In many ways, direct marketing is the perfect form of marketing for telecommunications companies to use.

First, the telcos (at least the incumbent telcos) already have the names, addresses, and phone numbers of every person in the world with whom they might possibly want to do business. Second, the telco (again, the incumbent telco at least) already has a good, personal relationship with most of these people. Third, the telco knows more about its customers than most other companies do.

Because of these three points, and because of the incredible efficiency and effectiveness of these techniques, telcos spend considerable portions of their budgets on this form of marketing.

7

Advertising and Promotion

It pays to advertise.
Anonymous

Advertising is the science of arresting the human intelligence long enough to get money from it.
Stephen Leacock

I don't care what you say about me, as long as you spell my name right.
George M. Cohan

In the previous chapter, we looked at the practices of direct marketing and saw how they fit into telecommunications marketing. In this chapter, we will provide a similar perspective regarding the use of advertising (broadcast media) and promotion.

The elusive nature of broadcast media and promotion

One of the most attractive things about the direct marketing approach is the remarkably tangible and measurable way in which it works. When it comes to the two other major forms of marketing, however, broadcast and promotion, this tangibility is quickly replaced by intuition, judgment, artistic flare, and nuance. Indeed, if advertising and promotion were not so incredibly successful and cost effective, most telcos would rather exist without them altogether.

Choosing whom to deliver the message to

While the discipline of direct marketing requires that the marketer work with lists as a starting point for all activities, the broadcast marketer has a much larger pallet from which to work. For the most part, the broadcast marketer deals with the entire population of a continent, country, region, or city as a starting point for the messages. Because of this, the whole process of preparing the message becomes especially important. While the direct marketer has the luxury of knowing who will see the message and, therefore, what the reaction might be, the broadcast marketer must develop a message that will be seen by both the intended audience and everyone else.

Since everyone will see the message, care must be taken that it sends the right communication to the people to be reached. Of course, to do that, marketers need to be aware of exactly whom they want to communicate with, and that requires a special discipline.

Determining whom to target the advertisements to

The process of determining who the customers are in your marketplace is known as *segmentation*. It is such an important part of direct marketing, broadcast marketing, and promotion that we spend three full chapters on it (see Chapters 10, 11, and 12).

Because we will spend so much time delving into the details of how it works, we will review the process from only the highest level at this point. Suffice it to say that segmentation defines for marketers exactly to whom they want to send a message, while at the same time providing clues as to how that message should be constructed for maximum impact.

Advertising development process

The process used to develop advertising is more loosely defined than it is for direct marketing. The marketer involved in the process of developing broadcast media-based campaigns usually goes through the following steps on the way to the actual publication of print ads or broadcast of electronic advertising messages.

1. *Objective setting.* Just as in the case of any marketing activity, the first thing is to determine exactly what they expect to accomplish with an ad. For the broadcast media and the advertising agencies that drive the process, these goals can be rather nonspecific and "visionary." Nonetheless, each marketer who puts a broadcast media message together must develop the same list of criteria as the developer of any other marketing activity.

2. *Segmentation.* Early in the advertising process, broadcast marketers spend a significant amount of time looking at segmentation data, especially that information having to do with the characteristics of the customers and prospects. While direct marketers look at characteristics-based segmentation information as a secondary aspect to the more critical behavioral information, broadcast marketers rely heavily on this information to determine to whom their messages must be directed and how to direct these messages.

3. *Message development.* Message development and the creation of powerful, high-impact marketing messages are the heart and soul of the advertising industry. During this phase, marketers create test messages. These *storyboards,* as they are called, whether they happen to be newspaper ad mockups or artist cartoon renderings to display the flow of a television ad, are used to promote the concepts for which the advertiser is trying to obtain funding.

4. *Message testing.* Often, after the initial approval of an ad, the marketing organization will sponsor test showings of the ads to gain confirmation of how effective they will be. It also gives them an opportunity to change the message before committing major resources to the execution of the campaign.

5. *Message delivery.* Finally, the advertisement is created and delivered to the customers.

6. *Feedback measurement.* Unfortunately, it is almost impossible for a company to measure the impact that a broadcast campaign has. The problem is that customers in the general marketplace can be subjected to many messages before they buy. Because of this, when a customer does buy, there is no way for the marketer to know which ad or combination of ads was responsible for that event.

Promotion development process

The development of promotional activities is even less disciplined than broadcast. In many cases, individual operational areas will decide that participation in promotional activities will contribute to a positive perception in the marketplace. For example, geographical area managers may choose to attend local county fairs, trade shows, and similar events. At the corporate level, upper management may decide that the firm needs representation at major global trade shows to promote the prestige of the firm. In all cases, the same criteria for marketing activity development, namely, the identification of message, market, and segment, are the foundation to the event.

Fitting advertising and promotion into the telecommunications marketing process

Just as in the direct marketing discipline, telecommunications companies apply these disciplines to their overall marketing process. Let's consider how this *fit* typically occurs in a telco.

Prioritization and goal setting

The prioritization and goal-setting process is an ongoing one. Because of this, it is easy to lose track of where the impetus for marketing activities comes from. One might assume that corporate objectives, operational demands, and long-range marketing plans are the key drivers to making this process work. In reality, the impetus for advertising and/or promotion is often instigated by two unlikely sources. We refer to these sources as *competitive activity* and *image building*.

Competitive activity

Many times, a telco will decide to participate in an advertising campaign or promotional activity for no other reason than that their competition does it. Usually, the thinking behind this decision-making process is "They are doing it, so it must work. We had better do it too!" or "They are doing it, and it makes us look bad if we don't do it too!" The irony, of course, is that this thinking is not necessarily bad logic. No matter what kind of rationalization is behind the telco's decision to enter the advertising and promotion competition with other telcos, ultimately telcos who advertise more will have better recognition. These telcos will attract more customers than companies that do not advertise.

The result is that two or more companies in the same market can create an advertising and promotion battlefield, where each escalates the marketing budget higher, and forces the entire industry to raise the bar to keep up.

Image advertising

The other reason that telcos might invest in advertisements and promotions (especially the expensive and "fancy" advertising) is because dynamic and powerful advertising agencies and internal marketing promoters convince executives that the *images* and *messages* they create will have a powerful and long lasting impact on the market's perception of the company. This kind of image advertising and promotion is very difficult to measure and explain within the context of the marketing process. It is prevalent nonetheless.

Strategy and advertising

While direct marketing tends to be chosen as a medium after some specific goals for marketing have been decided on, advertising and promotion tend to be part of the impetus for the overall process itself. The decision to participate in broadcast marketing is one with large strategic impacts, and the manner in which the level of advertising is established will have a significant impact on the success of the company.

When deciding if, when, and how to participate in the "big leagues" of the advertising world, the telco needs to decide on the basis of an understanding of its strategic position.

Deep pockets

If you are going to invest in significant advertising and promotion, you must be ready to back up that commitment with significant money. Advertising on television and radio in particular represents a huge investment for most telcos, and there is no guarantee associated with these ads.

Market share considerations

One of the first things for the marketer to consider in this form of marketing is market share position. This includes the following issues:

1. *What is our current share of the market?* A company that currently owns 10% of the market looks at advertising very differently than a company that owns 90% of the market. Mass marketing is an approach that favors the big players. The smaller your share, the less appealing this approach.

2. *How many other competitors are there?* The number of competitors in your market is also an important consideration. In a two-competitor market, advertising will have a significant direct impact. In a market with four or more competitors, however, advertising can do more to confuse consumers and create more of a "commodity view" of your market than you want.

3. *What percentage of the overall market do we want?* Mass marketing is best when your goal is to capture a significant portion of the

market. Mass marketing messages go to everyone; therefore, it is best used if you want to attract a large percentage of the people receiving the message. Otherwise, selective types of marketing are more appropriate.

4. *Is this the best way to attain that share?* Of course, the real question is whether or not you think the advertising approach will gain you the share for which you are aiming.

Well-funded versus lightly funded competitors

The decision to compete with mass marketing is one that favors the company that is well funded. Because we know that the tendency of advertising battles between competitors is to escalate the level of activity, the company with the most money available to support the battle will end up with the biggest and best message.

For this reason, a less well-funded competitor should consider minimizing mass marketing activities, if for no other reason than to reduce the chances that the better funded competitor will escalate, creating an environment in which it is even more difficult to compete.

The risk associated with weak commitment

The other risk a player in advertising takes on is that a low level of commitment to mass marketing can create a cumulatively negative impression of the company, making it look smaller than it actually is. The impact on the market of the cumulative effect of all messages needs to be taken into account before embarking on a major campaign.

Capabilities and infrastructure

The other point that must be considered is that any commitment to mass marketing must be made against the backdrop of the company's current strategic position, in terms of its network infrastructure, capacity, and customer support capability. Many a telco has invested in mass marketing efforts, only to generate more customers than it could support. This creates huge network, capacity, and customer service problems, resulting in a net negative impact.

Modeling

While strategic prioritization and goal setting is by far the largest component of a decision to move forward with an advertising activity, most organizations also conduct extensive modeling activities to back it up. These activities include the following:

- *Segmentation.* Learning which customers you want to work with and what motivates them to buy.

- *Focus groups.* Calling a group of customers together and asking them questions about their likes, dislikes, preferences, and impressions of ads, offers, and products.

- *Surveys.* Marketers often gather key data to feed their modeling efforts through the use of surveys, where customers are asked questions about themselves, their attitudes, and their desires.

Campaign planning

As the advertising or promotional message is polished and perfected and after the decision to run the campaign has been made, the marketer will usually run tests to validate the effectiveness assumptions that were developed during the modeling process. At the same time, all of the detailed logistics of message creation, including actors, studios, artists, and the myriad of other people involved in the creation of advertising or promotional support objects are employed and directed to the task of creating the real campaign.

Campaign execution

The campaign is executed, and the message is delivered to the mass market.

Feedback and measurement

Ultimately, the company is left with a need to measure the effect of advertising and promotional activities on their business. Directly measuring the impact may not be possible, but there are ways, either passively or through inference, to estimate the effects on business.

In future chapters, we will revisit the issue of tracking and measurement and see how the creation of a marketing database environment can make the measurement of effects possible.

8

Acquisition, Retention, and Wallet-Share Campaigns

The quickest way of ending a war is to lose it.
George Orwell

Wars begin in the minds of men.
UNESCO

Regardless of which media they choose, ultimately the marketers must get down to the difficult business of figuring out what they want to say to the prospects and trying to ensure that the customers react in the way that they had hoped for. This is the true art and science of marketing, or the development of wallet-share, acquisition, and retention campaigns, also known as *WAR*fare.

Variations in approach based on objectives and messages

The selection of a particular medium to use in the execution of a campaign will have significant effects on how the marketing database is built and how the marketing process is run. What has even more impact, however, is the objective that the marketer is trying to accomplish. To illustrate how different the process is in each case, we will walk through some scenarios.

Objectives and messages: what is the difference?

By now, you will have noticed a similarity between (1) our list of *objectives* that a marketing sponsor has for executing a campaign, (2) the types of *messages* that a marketer will be setting up to deliver to customers, and (3) the way we categorize campaigns themselves. We can describe all three by their common elements, these being wallet-share, acquisition, or retention criteria.

Acquisition campaign

Let's begin with an *acquisition* campaign and see how the different elements of the marketing process come into play. For a telecommunications firm, what are some of the forces that come into play that would prompt a marketer to conduct an acquisition campaign? The answer depends on a number of conditions.

Three major types of acquisition campaigns

An acquisition campaign is necessary for companies in three situations:

1. Companies that are new entrants into a given market (start-up campaigns);

2. Companies dealing in markets with an increasing number of prospects (incremental campaigns);

3. Companies interested in acquiring customers from a competitor (predatory campaigns).

In all three of these situations, the company is looking for new customers for obvious reasons.

Start-up campaigns

Start-up campaigns are some of the easiest and most straightforward to run. A new market is made available and the company needs to present itself to customers, signing up as many as possible.

The most common form of start-up campaign occurs when new telecommunications products are introduced to the market. Most recently we saw this happen with the plethora of wireless providers sweeping across their respective markets, selling first cellular and then also PCS service.

Just because the company is entering a new market does not mean that the discipline of a marketing process is no longer needed. Prioritization, modeling, development, and execution must still be approached with the same care as in more specific campaigns.

Incremental campaigns

Another acquisition case occurs when an existing telco needs to aggressively continue its acquisition momentum. As the presence of telecommunications products matures in a given marketplace, new groups of customers decide to take the plunge and sign up for a product they have never owned. Campaigns to seek out and appeal to these kinds of customers are similar to start-up campaigns, but must be more focused to be effective.

Predatory campaigns

The "steady-state" telecommunications firm will inevitably find itself in the position of wanting to increase its own market share at the expense of the competitor. Campaigns of this type take on a special flavor all their own, with some interesting twists on execution as they advance through the marketing process.

Acquisition: motive forces

These are the motive forces that drive a telco toward the decision to carry out an acquisition campaign:

- *Competition.* The presence of competitors and their level of acquisition activity provide marketers with a clue about their need for action.

- *Customer.* Market research about the demographics and characteristics of existing and newer customers can indicate to the marketer that untapped markets are available.

- *Product.* Existence of new products is the main reason to launch this kind of activity.

The well-managed telecommunications marketing process will provide upper management and the marketing staff with good, solid intelligence that enables them to see: (1) the current and future potential market for their products, (2) their current market share, and (3) the potential they can go after. The decision to run an acquisition campaign should be based on this kind of intelligence. It enables campaign sponsors to tell marketing not only that they want an acquisition campaign, but also specifically how many and what kind of customers they expect to acquire.

Acquisition: spoiler forces

Acquisition campaigns are directed and stalled by the typical spoiler forces. As is always the case in telecommunications, the regulatory environment will be the more critical spoiler force in effect. The marketer needs to be aware of what the past, present, and future regulatory environments will be and be ready to use that information for short- and long-term strategic advantage. Also important, especially for the acquisition campaign, is that the marketer understand the current and future capacity of the network and the operational systems.

Combining the motive and spoiler information, the marketer should be able to set a general direction and specific objectives for acquisition activity.

Now that the external and internal "messages" have come into play, the telco should probably look at running some kind of acquisition campaign. Let us turn now to the marketing process and see how these forces initiate and direct the company to react in an appropriate manner.

Acquisition: prioritization and goal setting

Assuming that the management of the telco is not asleep at the wheel, there will probably be considerable pressure from many people and organizations within the telco to *do something*. This general pressure and organizational consensus will become the impetus for the formation of a project team to act on that momentum.

Acquisition sponsorship

The sponsorship of an acquisition project will come from many different places. Upper management or marketing executives may make the decision that this kind of campaign is required. For a start-up company, for instance, development and execution of a start-up campaign is one of the largest single steps in the rollout of the company. In less dramatic situations, the decision to run incremental and predatory campaigns will be initiated by the people responsible for managing specific product lines, sales channels, or customer segments. In these cases, the operational managers will have noted an opportunity and decide that acquisition campaigns are appropriate and essential.

Acquisition teams

The team assembled to carry out an acquisition project includes people from the sponsor's organization and from marketing. Specific information about the potential target customer segments and the nature of successful past acquisition campaigns is helpful. In the case of major start-up campaigns, the services of the advertising agency that will be assisting with the rollout will be critical. In those cases where a direct marketing approach will be used, the vendor of database marketing services and lists will also be included.

Acquisition: modeling

Upon formation of the acquisition team, the project will quickly move into the modeling phase. While the modeling process has some common components shared by all campaigns, several things are unique to the way that acquisition modeling is done.

Preselection, postselection, and nonselection models

The first consideration of the team is whether the acquisition operation is to be a pre-, post-, or nonselection campaign.

A *preselection campaign* requires the marketing team to decide on the type of customer they want to attract and then shape the campaign accordingly. This is the only approach that requires a lot of modeling work. In this case, the marketing team undertakes the process of developing segmentation profiles of the customers they want to attract, and feeds that information to the campaign developers for focusing efforts.

In a *postselection campaign*, the marketing team decides that they would rather send out an attractive general offering to potential customers, and then allow the customer service and sales organization to postscreen the potential customers, deciding at that time whether to accept them or not. This approach is commonly used to screen out bad credit risk customers.

A *nonselection campaign* indicates that the telco will be completely indiscriminate about who they attract. In this case, any customer is considered a good customer. Unfortunately, many telcos run all their campaigns as nonselection campaigns. The net result is, of course, that they also attract many undesirable customers.

Acquisition cost models

Another way of looking at the process of acquisition is to determine the typical or average cost of acquiring a single customer and then using that data to determine how to size the overall campaign.

The calculation of the cost to acquire a new customer is a standard metric. This metric is defined by most telcos. It usually involves some variation on a basic process for determining the number of new customers acquired over a period of time, and dividing this number by the amount of the marketing budget for the same period. Armed with the basic acquisition cost information, the marketer can manipulate acquisition cost elasticity curves and more accurately size the acquisition campaign.

Cost of acquisition by channel model

Sophisticated telcos use the basic cost to acquire a customer model and broaden its applicability by comparing the cost of acquiring customers through different channels. For example, the marketer may find that television ads cost about $300 per acquired customer, while retail channels deliver them at $450 per customer, and direct marketing campaigns at $250 per customer. With this kind of information, the marketer can properly balance the different approaches.

Price for points model

The most common method used by mass marketers to calibrate the objectives and effectiveness of a marketing campaign in the traditional advertising world, is the use of what is called a *price for points model*. This model examines the demographic segments common to a given geographic market and assigns a dollar cost (price) to the percentage of the demographic segment (points) that will be reached. In other words, "I can reach 10% of the young urban professional market in Cincinnati with a television ad placed during the *Drew Carey* show for a price of $X.$"

The price for points model is used by advertisers to negotiate rates and measure effectiveness, as reported by media penetration reporting companies like A. C. Nielsen.

Lifetime value model

The *lifetime value model* requires the marketer to determine the long-term revenue potential of a customer type, and then use that information to target potential customers. This focusing introduces a far more manageable and measurable kind of accountability into the marketing management process.

Preferred behavior model: the customer value function

In Chapter 13, we discuss the customer value function in detail. It is a calculated value that provides the marketer with specific gradated information about the potential overall profitability, reliability, and potential lifetime value of a customer. It then combines this information into one metric, which can be used to drive all campaigns more precisely. The *preferred behavior campaign* is an approach that uses this customer value function to accomplish these objectives.

Hybrid approaches

Of course, there is no rule that says a marketer must use only one of these approaches. Commonly, marketers will combine these and others into their campaign approaches. Our chapter on modeling (Chapter 15) discusses this in detail.

Acquisition: campaign development and execution

Given the many different ways that an acquisition campaign will be focused and modeled, it should come as no surprise that there are some variations in development as well. Because of the nature of an acquisition campaign, certain things will be unique about its development.

Customers without history

Differing from wallet-share and retention campaigns, acquisition campaigns alone require the company to send messages and approach people with whom they have no relationship. This has varied consequences for the development of those campaigns.

In the case of direct marketing campaigns, the telco must obtain the list of names that they want to approach from a place other than their own records. Obviously, if we have a record about an individual, then by definition this person has already been acquired. For more information about lists and acquisition of names of people who are not customers, see Chapters 17 and 18, which discuss marketing databases and list management processes.

In the case of mass marketing campaigns, the fact that the people the telco wants to attract are not current customers can have a different twist. For a new start-up, running a start-up acquisition campaign will be like shooting fish in a barrel. Because nobody is a current customer, everybody can be approached. As soon as the company starts to get some customers, however, the mass marketer has to be sure that the messages being sent do not make existing customers unhappy. For example, if the telco runs an acquisition campaign that offers service at a discounted price, lower than current customers are paying, they run the risk of losing more customers than they gain. Therefore, in the majority of the cases, the mass marketer needs to be aware of the impact of acquisition messages on both prospects and customers, if the campaign is to be maximally effective.

Acquisition: feedback and measures

The early generations of telco marketers thought that acquiring feedback and measuring a campaign's effectiveness in the acquisition world were easy tasks. They assumed that keeping a simple count on new customers or activations was the only metric that was needed. This simplistic view, however, is quickly giving way to a much deeper understanding, which requires marketers to track and measure many different characteristics of their acquisition activities. Keeping track of who signs up for products is valuable information for marketers as they prepare to run more campaigns. In integrated, long-term, strategic marketing the marketer needs to record much detail about the acquisition process itself.

Retention (churn) campaign

While acquisition campaigns are relatively straightforward, retention campaigns tend to be somewhat tricky. In the acquisition case, the marketer's job is simple. Get more customers by hook or by crook! Of course, as every businessperson knows, and what every telco is learning, there is a difference between getting customers and keeping customers.

Retention campaign approaches

Just as there are different ways to approach the acquisition problem, there are also different ways to keep customers. The purpose of a retention campaign is to determine who is the most likely to leave and, in most cases, why will they leave. Then do something to dissuade them from leaving.

The big challenge with retention campaigns is problems with timing. To be effective, the retention campaign must determine who is likely to leave and why, and then make a change that will cause them to stay. This must be accomplished long before the people actually know they want to leave. So if the retention campaign is to be effective it has to provide these insights with enough lead time to actually address the problem.

Retention campaign approaches are discussed next.

Buyer reinforcement campaigns

Over several decades, the automotive industry learned that most buyers inevitably undergo a period of *buyer's remorse* shortly after buying a new car. This tendency has created all sorts of problems for companies. Over time, they have developed a marketing approach to address buyer's remorse and help reinforce buyers' decisions. The number one tool in the anti-buyer's remorse arsenal is campaigns geared toward making customers feel good about the decision they have made. These ads, though they may look like acquisition ads, are really designed to make recent car buyers feel that they made a good decision. Some telcos are starting to design similar postpurchase reinforcement-type campaigns.

Image enhancement/reinforcement campaigns

Along the same lines as the buyer reinforcement campaigns are the image enhancement campaigns. These campaigns are directed toward all current customers.

For the telco, the purpose of these campaigns is to convince customers to feel good about the telco as their service provider. Furthermore, the purpose is to cause them to feel guilt or remorse if they were to change service providers; to feel as if they would be losing something.

Price adjustment campaigns

Another way to prevent customers from leaving due to a lower competitor price is simply to lower your price. This approach can be sound, but only if you are sure:

- That customers are preparing to leave;
- That offering the reduced price will keep them;
- That you can afford to offer the service at that price.

Many telcos believe that price adjustment campaigns are the only way to address retention issues. In reality, an overly aggressive use of this approach will drive down profits and decrease customer confidence.

Service level adjustment campaigns

The other significant attempt a telco can make to convince customers to stay is to increase/improve the level of their customer service. Many times customers want to leave, not because of a competitor's lower price, but because of the better service they will receive, either in the quality of their lines, or in the quality of their relationship with the company.

The "devil you know" campaigns

Finally, there are campaigns that appeal to the customer's sense of loyalty and fear of the unknown. These campaigns basically issue a warning that says, "You don't want to change; you don't know what might happen with those strangers."

Retention and the spoiler and motive forces

While the telco's upper management, sales organization, and operational units are all quick to identify and demand that marketing do something about the need for new customers, the inspiration and motivation to address retention usually comes from very different sources.

Who cares about retention?

Because of the history of most telecommunications companies, they do not have a natural concern about customer happiness. If you thought it was hard to get a telecommunications executive to understand that paying big money for customer acquisition was challenging, try convincing the same executive to spend even more money to keep the acquired customers! The concept seems preposterous.

As unbelievable as that sounds, think about the origins of most telcos. Many have previously been protected monopolies with guaranteed markets. In most countries, the telco was doing you a favor if they gave you any service at all! Even in relatively advanced and proactive economies, like Brazil, the waiting lists for new phone lines take months or even years to fulfill. To ask companies with this cultural predisposition to jump to the other end of the spectrum, to proactively solicit customer opinions, and actually care about what the customer thinks, is a giant step.

The telecommunications company cultures conspire against the retention of customers in many ways, and one of the biggest is sales. For years, the sales of telecommunication services and products to customers was not measured on how much revenue customers brought in or how loyal those customers were. It was instead based on the number of new customers. The main and often only metric used to measure the effectiveness of acquisition campaigns and sales channels is typically new activations or the number of new customers added in the current month. Because of this, it is usually the customer service or accounting organizations that reveal to management that there are serious downward revenue and subscription trends that need to be addressed. These organizations become the force responsible for motivating the telco to seriously address retention issues.

Double and triple counting—the professional churner

One of the most unfortunate things about an activations-based form of sales measurement is that this metric encourages sales channels to seek out and pursue professional churners. The term *professional churner* describes individuals who willfully and habitually change carriers on a regular basis to take advantage of the bonus offerings that telcos make to new subscribers. Because most telcos only track activations by raw count, and do not track when or how individuals are acquired as customers, they pay repeatedly to reacquire the same customers.

Forces affecting retention

For the average telecommunications firm, retention problems are created in several different ways. From the motive force area we find that both competition and innovation play a key role: competition, because it is the competitors who give customers a place to churn to, and technological innovation because new telecommunications options can motivate customers to abandon older technologies for newer ones. Witness the mass migration to PCS from cellular.

The spoiler forces involved in the retention equation include the regulatory environment that allows or disallows competition, often defining how competition can be conducted, and the company's own network infrastructure, which limits or enables the telco's range of product offerings.

Retention: prioritization and goal setting

For a telco concerned with retention issues, the challenge is ensuring that there are proper internal and external reporting mechanisms to promote awareness of details regarding the company's retention profile.

The retention profile

We use the term *retention profile* to describe the current status of the telco's customer base, measured in terms of how quickly that base is "turning over." In other words, when a telco has a stable and growing population of customers, we say that their profile is a good one. However, when the number of customers is declining, or many customers are leaving and are replaced by new customers, we say that the profile is poor.

All companies and for that matter all telcos have a certain amount of turnover. This is a new experience for most telcos, since they are accustomed to being the only supplier in the market. For telcos, there were only two ways to lose a customer (when they died or when they moved).

When to run a retention campaign

Gaining and losing customers is a natural part of any business; so how does the telecommunications company decide when a retention-type campaign is in order? Unfortunately, for most telcos the answer is to wait until the number of subscribers begins to fall so dramatically that there is fear for the future viability of the firm. What would seem to be the more appropriate long-term trend is for telcos to monitor their retention profile regularly and begin making proactive decisions about when and where to initiate retention activities. The test for the efficiency of the marketing department in this case is based on their ability to foresee retention problems and preemptively run campaigns to address those problems. After all, it is easier to keep customers loyal than it is to win them back.

Retention: modeling

The job of determining how to develop models for the retention case is an interesting one. People use many different approaches to develop models in supporting the various retention campaigns (buyer reinforcement, image enhancement, price adjustment, service improvement, and "devil you know").

Remember that there are three parts to the definition of these models: (1) determining who is going to leave, (2) determining why they will leave, and (3) determining what to do to get them to stay. Some of the modeling approaches that can help are discussed next.

The retention cost versus acquisition cost model

An organization concerned with retention issues can calculate (1) costs to acquire a new customer versus (2) costs to reacquire a former customer versus (3) costs to search for a completely new customer.

Price elasticity models

Another approach that can help the modeler get a feel for the situation, especially where price issues are concerned, is to compile price elasticity models that indicate how many customers will be added or deleted for each change in the price of the product. These elasticity models make it possible for the marketer to accurately evaluate outcomes as different pricing levels are attained.

Attrition behavior models

While retention cost and price elasticity models provide the marketer with general directions to follow in the retention case, the genuine hardcore value will be found in the development of attrition behavior models. These models describe for the telco what the specific utilization, buying, and profitability behaviors are for each customer. This allows telcos to focus on the most valuable customers.

Attrition value functions

Another variation on this theme is to create a customer attrition value function that provides a numeric indicator, illustrating how likely individuals are to leave and how much their departure would cost the telco.

Retention: campaign development and execution

The preparation and execution of retention campaigns can be even more challenging than the acquisition campaign. In the case of the direct marketing scenario, the marketer will not need to use external data since the concern is only with keeping current customers. On the other hand, external information might be purchased, enhancing the marketer's understanding of who the customers actually are.

Retention campaigns that utilize mass marketing are especially tricky. The marketer ensures that the message being delivered does not unsell, undervalue, or overvalue the company or the service. This considers the perceptions of consumers that the telco is not trying to retain, but who might also receive the message. Imagine the impact on sales if the telco ran a mass marketing ad that offered to reduce new customers' prices. It could very well cause existing customers to become discontent.

Retention: feedback and measures

While the development of feedback and measures is helpful for acquisition campaigns, it is absolutely crucial for running retention campaigns. How can you possibly know who is or is not leaving if you do not report on such activities? How can you build the models you need and develop the conclusions that are required for the development of good campaigns? Certain key metrics are fundamental to the successful execution of any retention-related activity.

Wallet-share campaign

Whereas acquisition and retention campaigns are the main business of wireless and long-distance companies, wallet share is something that the wireline providers are currently working on aggressively. The main objective of a wallet-share campaign is to convince current customers to spend more money than they have in the past.

Types of wallet-share campaigns

Coming up with a good inventory of wallet-share campaign types is difficult because so many telcos are in such different positions and, therefore, many types of approaches are possible. Some of the more popular approaches follow.

Bundling

Bundling is the telecommunications industry practice of identifying related products and services and selling those collections as single-priced offerings. Bundling takes many forms. A wireline carrier might bundle features such as three-way calling, call forwarding, and call waiting into deluxe packages. This encourages customers to buy more items.

A multiline vendor like AT&T might bundle complete lines of business, such as wireless, Internet, long-distance, and pager service, into one flat-rate package to gain overall market share and create a loyal consumer base for future expansion.

In all cases, the key to good bundling strategies is to understand the costs, revenues, and net gains leveraged by bundling activities.

Bundling as an acquisition and retention strategy

While bundling is first and foremost an approach to maximize profits for a given group of products, many clever telcos have realized that bundling can also be used to attract and retain customers. The customer's ability to engage a single provider for all telco services has its advantages, such as one bill and one block, discounted price. This can provide a compelling reason for customers to join the company, stay with it, and increase the wallet share.

Consumption enhancement

The second type of campaign activity is *consumption enhancement*. In this case, the marketer looks at the customer's current level of service utilization and tries to determine how this consumption might be increased.

Portfolio expansion

Another very popular type of wallet-share campaign is *portfolio expansion*. These campaigns often involve incentives or discounts to encourage customers to expand their current portfolio.

Wallet share: forces

The forces at work that move the telco to take on a wallet-share campaign are very different than those in the other two cases. Most telcos do not aggressively pursue these options until after retention or acquisition is no longer critical. A company will do wallet-share work when the customer base is relatively stable. Obviously, if you are experiencing 50% churn or are trying to acquire thousands of customers a month, your focus will be on more profitable kinds of activities.

Eventually, as the dust settles and parts of your market stabilize, the need to increase the revenue per customer arises. In this case, it is the absence of most of the marketing forces that creates the optimum environment for wallet-share activity. These campaigns are viable when (1) competition and innovation are stable, (2) no large regulatory changes are on the horizon, and (3) the company's network and operational infrastructure can support such a campaign.

Wallet share: prioritization and goal setting

How then does the marketing department decide that wallet-share campaigns are a good idea? That depends on the situation. It will likely be upper management, looking at the overall "big numbers," that notices pockets of opportunity that are not being addressed. At the same time, many operational managers will realize that their own products and channels can benefit from the accomplishment of their objectives, through the creation of "partnered" campaigns.

Wallet share: modeling

The job of modeling is especially complex, challenging, and interesting for the marketing team. In this case, information about detailed customer utilization behaviors, buying habits, and attitudes combine with detailed product cost and profitability information to create a rich set of supporting model templates.

Market basket analysis

One of the most popular forms of analysis for developing bundling scenarios is what retailers know as *market basket analysis*. Market basket analysis examines the various items that consumers put into their shopping baskets in a single trip to the store. The resulting data provide information about how to better price and market these related products. Market basket analysis might show, for example, that 50% of the people who buy cookies also purchase milk. Such information allows the marketer to tie the sales of the products together more profitably.

In the same manner, telco marketers can analyze the kinds of products that consumers buy at the same time: Cellular service and pager service? Call waiting with call forwarding? Different associations will give marketers clues to better bundling strategies.

Dragger/draggee analysis

Dragger/draggee analysis is a specific kind of market basket analysis. It figures out which products will draw people in and which will be bought as a side effect. For example, the sale of cameras will drag the sale of film. The sale of electronic toys will drag the sale of batteries. Marketers can identify similar dragger/draggee relationships between telco products and build wallet-share programs accordingly.

Utilization behavior analysis

Along with the straightforward market basket approaches, modelers can also examine utilization and call detail information to determine specific patterns to exploit.

Wallet share: campaign development and execution

Wallet-share campaigns bring with them issues similar to those experienced by the developer of retention campaigns. The fact that all prospects are existing customers, and that the messages sent must be specific to be effective, leads us to the same sets of constraints.

Wallet share: feedback and measures

The key to the success of these campaigns, more than any others, is the ability of the marketer and the model to work with information about product profitability and price (especially the bundled price). It must also take into account the elasticity that customers will tolerate or embrace in response to different schemes. The wallet-share campaign is the most sophisticated and difficult to measure, but often it is the most profitable.

P a r t 3

Analytics

9

Product, Customer, and Competitive Analysis

Knowledge is power.
Anonymous

In the second chapter, we introduced the concept of the marketing planning and execution cycle. That cycle describes, at a high level, how to participate in the process of marketing for any given company. Although this high-level view of both the process and objectives of marketing gives us an idea of what a marketer does, it does not help us understand specifically how a telecommunications firm actually makes these processes work. What is clear, however, is that throughout the execution of the various parts of the marketing process, there is a core set of analytical information that must always be available if the marketers' decisions are to be fact based and accurate.

Three general areas of investigation and analysis support the marketer throughout the process of strategy development, prioritization and goal setting, modeling, campaign development, and campaign execution. These three types of analysis allow marketers to understand in tangible and quantitative terms exactly what they are selling, who is buying, and what the competition is doing about it.

Product (line-of-business) analysis for telecommunications

Perhaps one of the most difficult parts of the marketing cycle for telecommunications firms to deal with is this area of analyzing the different products and services that are offered from within an analytical framework. Product analysis (also referred to as *line-of-business analysis*) is something that telecommunications companies have only started to worry about. Other industries (like retail and manufacturing) have always known how critical the understanding of a product's profitability and market position are to the overall health of the company.

Telecommunications firms, on the other hand, have only recently started to see any value in doing this kind of analysis. Remember, telecommunications firms used to have only one product to offer and, therefore, had no need for product analysis.

Today's telecommunications world is one in which the company is literally overwhelmed with alternatives. Engineers have hundreds of options to choose from in putting a network together, and companies have dozens of different products that they can consider offering in a variety of different ways.

(Throughout the balance of this book, we use the term *product* in its most general and generic sense, referring to all products and services that telecommunications firms offer to consumers.)

Product lines (line-of-business) inventory analysis

As trivial as it may seem, the first thing any developer of product line analysis within a telco must do is to develop an inventory of exactly what products the company is offering to customers. Most efforts to round up this kind of information result in some pretty interesting discussions. You see, the concept of a product line is not part of the vernacular of a typical telecommunications person, and so different people think of different things as products.

To some individuals, each separate "call pak" or "packaged service offering" that the firm makes to the consumer might be considered a product. To others, the product lines are defined as the basic services offered, such as local, long distance, pager, and cellular.

In all cases, it is critical that a single, unifying scheme be developed for the identification of these products and their organization into a hierarchy of product lines. Without this kind of hierarchical identification structure, it will be impossible to measure, monitor, and plan for the sales and maximization of a firm's revenue potential. Table 9.1 shows a typical list of product lines and products within those lines that a telco might offer to customers. Notice how the general category of "product line" includes both purely service offerings (such as local and long-distance service), pure product offerings (such as cellular and PCS phone devices), and combinations of the two.

Table 9.1
Product Hierarchy: Product Lines Inventory for the ABC Telephone Company

Product Line (Line of Business)	Product/Services Offering
Local phone	Local service
	Misc. support services
Wireless	Wireless devices
	Cellular services
	PCS service
Pager	Paging devices
	Paging service
Internet	Internet access
	Internet hosting
Voice mail	Voice mail service
Calling card	Calling card service
Long distance	Long-distance service

Product profitability

Of course, the development of a product line hierarchical scheme in and of itself provides little value to the company. However, putting such a scheme in place enables the company to take the next step in the process, understanding product and service offering profitability.

To put together any kind of marketing strategy, you need to know how much it costs to deliver a product to a customer or how much profit you make for each minute of its use. Without this, there is no way to figure out whether you want to sell more or less of it.

Here again, we have a situation where the telecommunications company is ill equipped to deal with the issues that this function brings up. For decades, telcos were able to exist without having any clear idea of what it cost to deliver its services to individual customers. (The telco was expected to deliver equivalent service to everyone, and the cost was evenly distributed across all of the customers.) However, in today's marketplace, understanding costs and profits are critical to good marketing planning. Most companies need to undergo some serious reengineering

Table 9.2
Product Line Revenues: Product Lines Inventory for
ABC Telephone Company

Product Line (Line of Business)	Product/Service Offering	Annual Revenue ($Million)
Local phone	Local service	$25.20
	Misc. support services	$0.78
Wireless	Wireless devices	$4.00
	Cellular service	$12.00
	PCS service	$7.00
Pager	Paging devices	$0.01
	Paging service	$0.70
Internet	Internet access	$0.56
	Internet hosting	$0.03
Voice mail	Voice mail service	$0.01
Calling card	Calling card service	$0.94
Long distance	Long-distance service	$15.43

efforts to get their hands on this kind of information. Table 9.2 shows the revenues for each product line for the ABC Telephone Company.

By examining Table 9.2, you might jump to certain conclusions about how well the ABC company is doing and which lines of business it should be concentrating on. But let's see how cost and profitability information

Table 9.3
Alternative Profitability Scenarios: Product Lines Inventory for ABC Telephone Company

Product Line (Line of Business)	Product / Service Offering	Annual Revenue ($Million)	Costs: Version 1 ($Million)	Costs: Version 2 ($Million)	Profits: Version 1 ($Million)	Profits: Version 2 ($Million)
Local phone	Local service	$25.20	$12.70	$22.90	$12.50	$2.30
	Misc. support services	$0.78	$0.60	$0.10	$0.18	$0.68
Wireless	Wireless devices	$4.00	$3.00	$1.60	$1.00	$2.40
	Cellular service	$12.00	$3.89	$11.80	$8.11	$0.20
	PCS service	$7.00	$2.57	$6.40	$4.43	$0.60
Pager	Paging devices	$0.01	$0.01	$0.01	$0.01	$0.01
	Paging service	$0.70	$0.39	$0.39	$0.31	$0.31
Internet	Internet access	$0.56	$0.25	$0.10	$0.31	$0.46
	Internet hosting	$0.03	$0.03	$0.03	$0.00	$0.00
Voice mail	Voice mail service	$0.01	$0.01	$0.01	$0.00	$0.00
Calling card	Calling card service	$0.94	$0.50	$0.43	$0.44	$0.51
Long distance	Long-distance service	$15.43	$12.40	$14.30	$3.03	$1.13

change that perspective. Table 9.3 shows two alternative cost scenarios and the resulting alternative profitability numbers for each line of business. How does the alternative profitability information change your inclination to spend money marketing to customers in different product line areas?

Of course, we do not want to be making decisions about when and where to invest the company's marketing dollars based solely on a few simple product profitability reports. Too many other options need to be considered. What these reports provide us with is a starting point, a collection of clues, if you will, that can help us decide where further investigation might be in order.

Consider the following: First, we may be able to impact these profitability numbers by examining our pricing schemes. (See the next section on product pricing for more information about this alternative.)

Second, we might also impact the profitability of a product area by reducing the costs associated with the delivery of the product or service to the customer.

Third, the fact that a product area is not very profitable does not especially mean that we need to do anything at all. Companies will often use unprofitable product offerings to help attract and hold customers who then spend money on the more profitable lines. Of course, the only way to know if this is a good idea is to gather much more information about our customers and their buying and spending patterns. This information is obtained through the execution of customer segmentation analysis.

Fourth, without information about what the current and future size of the market for that product might be, there is no way for us to know if these numbers are reasonable or not. We cannot judge whether we are simply paying our dues to earn a position in a market that will be extremely lucrative in the future, or whether we are supporting a product line that customers are not interested in anymore. To make that assessment, we need to look at competitive analysis and product life cycle issues.

In other words, product profitability information gives us a starting point for our analysis of the business, but does not provide enough information to make intelligent strategic decisions.

Product pricing

Another new and interesting discipline that has recently come under the explicit control of many telecommunications firms is the ability to set prices on a customer-by-customer basis. In the past, telcos were severely limited in their ability to manipulate pricing decisions because of legal and regulatory constraints. The telco had to offer the same rates to everyone, and any attempt to change those rates was handled through a complicated process, driven by the need to comply with a myriad of governmental dictates and rate schedules. Today, those requirements may still exist in some situations, but more and more, telcos are allowed to set their own prices. In fact, some companies are quickly moving toward the situation that the airlines are in, where every customer negotiates a unique rate for each service.

The art and science of developing pricing strategies is an interesting and complicated one, requiring the extensive use of statistical analysis and access to much historical and behavioral information. As we continue to develop our model for running a telecommunications marketing strategy, we will see how all these pieces can be fit together.

Product research and development

The forward thinking telecommunications firm is also always contemplating new products to be delivered in the future. Cable phone services, wireless Internet access, and cellular phones that function as pagers and computer terminals are just a small sampling of the dizzying array of new product options that the firm must consider to stay competitive.

Customer analysis (segmentation) for the telco

Understanding what products you are offering and how much money you can make with them is only the first step in the marketing process. The more important tasks revolve around trying to understand who your customers are and what kinds of products they want. This process, identifying who customers are, what they have bought in the past, and what they are the most likely to buy in the future, is known as *segmentation analysis*.

If you want to have a confused conversation with a group of people at a telecommunications firm, just ask them about the segmentation that they use. Talk to three different people and you are just about guaranteed to get three different answers. The problems, challenges, and uses of segmentation are numerous and complex and much concentrated effort is needed to sort them all out.

What is segmentation?

Segmentation, in the simplest sense, is the process of developing different schemes for categorizing and organizing groups of customers.

There are many different kinds of segmentation schemes, all of which are important to the development of marketing strategies. You can segment your customers by age, by geographical location, or by the products they buy. You can segment them by demographics, by psychographics, by simple criteria, or by using advanced statistical analysis techniques. Ultimately, however, the objective of the segmentation process is always the same: Segments are developed to help the organization understand who their customers are and what they want.

As we discussed in Chapter 2, segmentation is a specialized form of modeling and one of the most critical parts of the overall marketing modeling process.

How do you develop a segmentation scheme?

The process of segmentation is often referred to with a certain amount of hushed reverence by some of the more serious marketing practitioners. In reality, however, segmentation can be very simple and straightforward.

Segmentation schemes can be developed in several ways. Some are very simplistic; some are very complex, like the more "glamorous" approaches of advanced statistical analysis or data mining. Ultimately, though, regardless of which method was used to get there, the objective of the segmentation model is to simplify and clarify our understanding of who our customers are and what they want. In the next few chapters we investigate these methods of segmentation (and others) in much greater detail.

No matter how much energy you put into it, ultimately, a segmentation scheme tells you that you have different kinds of customers who behave and react differently to different situations.

Table 9.4 shows three of the more popular types of segmentation schemes. Customers can be segmented by their income, age, or psychographic profile.

What do you do with a segmentation scheme?

Segmentation analysis provides the foundations with which organizations run successful marketing campaigns. But what is so important about differentiating groups of customers?

Segmentation, like product analysis, plays a role in every part of the telecommunications marketing cycle. It provides the input that determines how the corporation should be organized and it contributes the data for the development of strategies. Segmentation helps markets set priorities, and helps drive the modeling process. It is a key component of campaign planning and execution, and it provides the yardstick for measuring how effective marketing activities have been. In other words, segmentation is the lifeblood that makes marketing work.

Table 9.4
Alternative Segmentation Schemes for ABC Telephone Company

Income	Age	Psychographics
Less than $35,000	Less than 16 years old	Young unemployed, no children (YUNC)
$35,001–$55,000	16–21 years old	Dual income, no children (DINC)
$55,001–$75,000	21–30 years old	Single-parent urban dwelling (SPUD)
More than $75,001	30–65 years old	Single-income, no children (SINC)
	More than 65 years old	

Associating revenue with segments

The easiest way to examine the power of a segmentation scheme is to associate the different categories with revenue numbers. For example, Table 9.5 shows us how much revenue was generated by each of ABC's particular segmentation schemes. Just as in the case of the product analysis, once we assign values to the different segments, we get a much clearer idea of how well our business is running and which customers are responsible for the success.

Why develop so many different segmentation schemes?

It is easy to see that the company needs some kind of separation of customers into different categories, but why develop so many different schemes to get the job done?

Couldn't we, for example, divide customers by age and sex and then use that scheme to measure everything? That would give us four separate market segments to base all analyses on: females under 18, females between 18 and 25, females between 25 and 50, and females over 50.

Although this simplified segmentation scheme may be helpful in some situations, we can attain much more information if we divide the market

Table 9.5
Revenue Generated by Different Segments of the Market for
ABC Telephone Company

Less than $35,000	$12.75	Less than 16 years old	$10.16	Young unemployed, no children (YUNC)	$0.30
$35,001–$55,000	$15.70	16–21 years old	$22.50	Dual-income, no children (DINC)	$105.45
$55,001–$75,000	$67.30	21–65 years old	$125.65	Single parent urban dwelling (SPUD)	$12.45
More than $75,001	$78.32	More than 65 years old	$15.76	Single-income, no children (SINC)	$55.87
Total Revenue	$174.07		$174.07		$174.07

differently. Segmenting customers by their income level or by the number of minutes of service used, for instance, would provide useful insights.

While we can use segmentation schemes to help us make all kinds of decisions in the running of the business, ultimately, we judge the value of a segmentation scheme on how well it allows us to understand and predict the behavior of customers in different situations, so that we can then maximize our profits. Therefore, segmentation helps the telco determine how to predict customer behavior and, thus, to maximize profit.

The better the segmentation scheme, the better able we are to predict people's buying behaviors. A good marketing department is always trying to figure out how to make their segmentation schemes better, because even the smallest incremental increase in revenues from a market segment can yield incredible increases in a company's profitability. The better the segmentation scheme, the better the company will be at meeting the needs of its customers, and increasing its efficiency and profitability.

How does segmentation help drive profits?

To see how a better segmentation scheme can mean success for a company, let's look at how helpful a couple of different segmentation schemes really are. Assume, for example, that we did a segmentation analysis by age and that it revealed that we have just about the same number of customers in every age group, and that each of these groups spends about the same amount of money. Has this segmentation analysis provided us with any useful information? Probably not. The segmentation by age exercise is not helping our marketing efforts in this case.

Now let's say that we segment our market by usage patterns. In other words, we would like to find out if there are different groups of customers who use our services differently. Our analysis might reveal that there are three types of customers: heavy users, who make more than 600 minutes worth of calls in a month; medium users, who spend between 100 and 600 minutes on the phone each month; and the light users, who use the phone for less than 100 minutes per month. Is this segmentation more useful in understanding our customers? Absolutely. This segmentation scheme shows that there is a select group of individuals who are heavy users of our service. Once we have identified who those people are, we can use the information in several ways.

1. We can go out of our way to make sure that these customers stay happy with the service (making sure that we do not lose our best customers).

2. We can offer special high-volume discount packages that will encourage these customers to use the telephone even more often.

3. Most importantly, we can study what kinds of people make up this high-volume group of customers, and begin an intensive marketing campaign to identify and convince more people like them to get their telephone service from our firm.

Segmentation analysis is an ongoing process

Segmentation analysis and the development of a better and better understanding of the customers is an ongoing process on which the marketing organization is constantly trying to improve. The key to the usefulness of a segmentation scheme is directly related to the usefulness of the information that it creates. If a segmentation approach allows the company to improve customer service, improve market share, reduce costs, or increase efficiencies in some way, then it will yield financial benefits that far exceed their costs. This explains why companies spend so much time and effort on segmentation analysis and why we will spend as much time discussing it.

Telecommunications competitive analysis

Clearly, understanding our own customers and our own products is an important ingredient in the development of a marketing strategy mix. However, when it comes to figuring out how we can get even better, it is helpful to understand how well our firm is doing compared to the industry overall and compared to specific competitors. Comparing our progress to the progress others are making can provide valuable insight into the nature of our successes or failures.

Industry trends and directions

One of the most interesting ways to gain insights into how well our company is really doing is to compare our progress to the industry overall.

Industry or government organizations often publish quarterly or annual reports that detail how many minutes of service were utilized by consumers, how much revenue was generated by the industry at large, and many other interesting facts about particular market segments and geographical areas. Comparison of our own organization's progress to these numbers can often prove to be insightful. Some companies have found that what they thought was an astronomical growth year turned out to be a year of less than impressive returns when compared to the industry overall. Others have found exactly the opposite.

In all cases, comparisons to the trends can help provide answers to the questions "How good of a job are we doing?" and "What kinds of goals should we be setting for next year?"

Product life cycle analysis

Another interesting set of insights and useful guidelines can be obtained by analyzing how well a particular product line has done against the backdrop of its consumer acceptance life cycle.

"It's not just plain old phone service anymore"

When the business of telecommunications was nothing more than simple phone service, the phone company did not have to worry about the popularity of its products. Phone service was phone service, and consumers wanted as much as they could get. Nowadays, however, with alternative forms of telecommunications available, telcos need to look at something they have never had to consider before: the obsolescence of some of their products.

Telecommunications products and services today are beginning to present the same kinds of consumer marketplace behavior that many of the other, more fashion-sensitive or technologically sophisticated products have exhibited for some time already. Phone companies are now offering products with marketplace life cycles that can be measured in months or years, compared to the old-fashioned "staples" of the industry—local and long-distance service—which had been operating on life cycles measured in decades or centuries.

Figure 9.1 shows a typical new product assimilation life cycle chart. It displays the typical pattern that consumers follow when new tech-

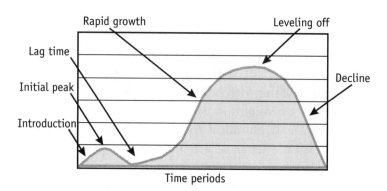

Figure 9.1   New technology assimilation life cycle.

nologies are presented to the market. As can be seen in this diagram, assimilation of new products tends to follow the same trend whether the product is a new type of car, a new computer, or a new telephone service.

In general the product will go through these phases:

1. *Introduction.* When a product is first introduced, most people may be reluctant to buy it, but certain "experimenters" and "innovators" will embrace it.

2. *Initial peak.* The momentum created by the innovators picks up and an initial peak of interest occurs.

3. *Lag time.* Suddenly, interest seems to disappear. The innovators are now bored with their new toys, and the full market itself has yet to decide whether or not it will embrace the technology.

4. *Rapid growth.* During this period, the market explodes with interest and sales increase rapidly.

5. *Level off.* Eventually, the market becomes saturated and equilibrium is reached.

6. *Decline.* New technologies come along, people's needs change, and the product loses buyers' interest.

Impacts of the new product assimilation life cycle on telecommunications marketing

Telcos need to plan for the marketing of these products in much the same way marketers of other short life cycle products do. Marketing planners need to have an understanding of the new product life cycle and assess where their products are in relation to that cycle.

Cellular versus PCS: the race toward obsolescence

Probably no area of telecommunications illustrates the problem of obsolescence management more clearly than the most recent experiences many companies have had with alternative forms of wireless service.

Within the past decade, the use of wireless telecommunications services has gone from being the exclusive plaything of a few wealthy individuals to a near necessity of modern life. From the streets of New York to the streets of Tokyo, it seems that almost every man, woman, and child is walking around with a mobile phone.

But of course, mobile phones, as a category, can be many things. The earliest major wave of mobile phone service was dominated by the cellular telephone. However, recent technological improvements have led to the introduction of digital versus analog, PCS, and even satellite communications systems, all of which are vying for a share of the wireless communications market (Figure 9.2). While the actual battle between cellular and PCS has only recently been engaged, significant evidence indicates that the appearance on the scene of the newer PCS technology has brought about the premature demise of the older form of cellular wireless communication, as indicated by Figure 9.2.

As the newer PCS technology takes hold, fewer people remain interested in the older technology and the subscriber base shrinks. When putting together a strategy for the exploitation of the wireless market, the marketer needs to decide which phone technologies and services the telecommunications company should pursue and when. Newer technologies are often less expensive and more appealing to consumers, but they require a sizable infrastructure investment to get them started, and there is no guarantee that the consumer will actually embrace the technology until long after that infrastructure has been built. Older technologies, on

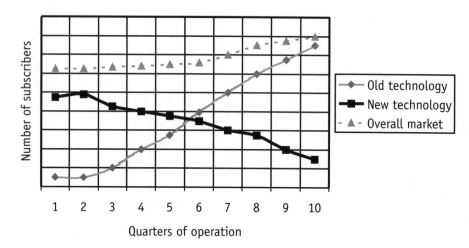

Figure 9.2   Cellular versus PCS.

the other hand, are less appealing to consumers, but are easier to sell and support.

The challenge then, is one of timing and risk management. At what point does the consumer's acceptance of the older technology stop and they are no longer willing to use them at any cost? At what point is it no longer profitable for the company to promote the different solutions? And how does the marketing department evaluate the best, most profitable decisions at each unique point in time? These kinds of problems are addressed through the use of product life cycle planning and analysis.

10

Simple Segmentation

You have but to know an object by its proper name, for it to lose its dangerous magic.

Elias Canetti

He brought them to the man to see what he would name them; and whatever the man called each living creature, that was its name.

Genesis 2:19

Naming things. Putting things into categories. This is one of the most basic and powerful mental exercises that human beings can exercise in the world.

The ancient Greeks believed that giving names to everything in the world, the elements, the animals, the plants, and the people, was critical to the creation of an atmosphere of order and peace. Many native peoples believed that by knowing the true name of someone or something, you could exercise magical control over it.

Linguistic scientists have proved that the number, type, and sophistication of the names you have for things affects the very way that your brain perceives the world. For example, Eskimo tribes often have more than 100 different words to describe the many different kinds of snow that they experience. The Eskimo does not see snow at all, but 100 very different forms of precipitation.

While our attitudes about the role that naming things plays in our modern lives may seem more sophisticated than those of our ancestors, the reality is that naming and categorizing things still plays a very important role in organizing our work environment and specifically in running a marketing organization.

Segmentation: an overview

Segmentation, the process of categorizing, understanding, and giving names to different groups of customers based on different sets of criteria, is one of the most important functions that marketing performs in business and even in our society itself.

How has our perception of ourselves and our society changed, armed with labels like Yuppie, Baby Boomer, Generation X, The Milleniers, and DINCs? What description of the twentieth century business world would be complete without visions of Madison Avenue executives spending tens of millions of dollars on advertising during the Superbowl games.

It is through the process of segmentation that the company learns to look at its customers in different, more accurate, and more realistic ways. Segmentation, in a very real sense, creates the lens through which the company looks at and focuses on the different characteristics and needs of that vast ocean of nameless, faceless individuals who eventually become trusted patrons of the firm.

Segmentation: a basically simple process

In the previous chapter, we considered a few of the different ways in which segmentation is accomplished. We talked vaguely about how marketers will sometimes use statistics and sophisticated data mining techniques to figure out how to put different segments together.

This super-sophisticated type of segmentation is not really what segmentation is all about, however. In fact, only a small percentage of the segmentation activities of most firms actually involve the use of these techniques. No, the vast majority of segmentation is based on nothing more than a fundamental understanding of who customers are and how our business works.

In this and the next two chapters, we will spend a considerable amount of time examining the issues involved in the development of segmentation schemes. In this chapter, we consider the overall process and the different types of segmentations, and also look at some of the simpler types, the structural and simple (univariate) categorical types.

In the next chapter, we will look at some of the more advanced segmentation types and techniques including the multivariate, multidimensional types, and look at how marketers develop these. In the chapter following that we will examine the process of segmenting based on behavior.

Segmentation: a formal definition

To start then, let's arm ourselves with a sound definition of just what segmentation really is.

The dictionary defines segmentation as "the process of dividing something into segments." It further defines a segment as "one of the parts into which something can be divided." So, literally speaking, we perform segmentation every time we look at something and try to figure out how it can be logically divided into different parts.

Customer segmentation, then, is the process of looking at our customers and prospects and trying to figure out how they can logically be divided into different kinds of groups.

Understanding segments and segmentation

Segmentation, while simplistic in its objectives, can be incredibly complex in its applications and approaches. In fact, to really understand segmentation, what it is, how it is done, and how it is used, we need to study it from at least three different perspectives.

We will, therefore, look at segmentation and the use of segments in terms of:

- The roles that the segments will play in each part of the marketing process (organizational, informational, predictive);

- The types of segments that can be created (structural, categorical, behavioral);

- The techniques used to define and discover segments (characteristics, classical statistics, neoclassical statistics).

Role of segments in the marketing process

A detailed examination of the marketing process, as we have described it, quickly reveals how crucial the development of accurate, informative segmentation schemes is to the execution of the marketing functions. In fact, the segments play three major roles in the running of the telecommunications business:

1. *Organizational.* Segmentation can provide us with valuable insight into the efficiency and appropriateness of the current organizational structure. Analysis of the different groups of customers serviced by the company often reveals that the creation of special business units, dedicated to meeting the needs of those groups of people, can often result in more sales, more profits, and more satisfied customers.

2. *Informational.* One of the most common uses of segmentation studies is to provide detailed information about customers to those people who need to know who they are and what they are like.

3. *Predictive.* By far the most potent reason to create segmentation schemes is to support the process of creating predictive models, which let us know how different groups of customers will behave in different situations.

Different types of segments

The marketer will work with basically three types of segments. Some of them are so obvious or straightforward that it requires no work to define them. Others are so complex and obscure that the marketer needs to spend much time and effort trying to figure them out. In all cases, the segments aim to help support the efforts of the marketer to better understand and simplify the company's understanding of how to better serve its customer base. The major categories of segment are (1) structural segments, (2) categorical segments, and (3) behavioral segments. A complete review of the different types of segments can be seen in Figure 10.1.

In this chapter we discuss the nature of and techniques used to develop and report on structural segments, and begin our exploration of categorical segments. In the next chapter we focus on the different types of categorical segments. The chapter after that is dedicated entirely to understanding behavioral segments.

Structural segments

Structural segments (also referred to as *natural* segments) are groupings of customers that are so basic and intrinsic to the nature of the business that people often don't even recognize them as segments. However, the fact that a segment is simple and basic does not mean that it is not a segment, created by the company, and consciously chosen to ease the management of customer relationships.

The three major categories of structural segments are geographical, product, and commercial.

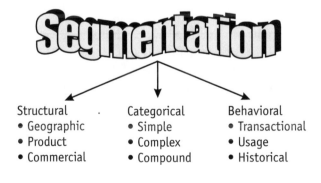

Figure 10.1   Types of segments.

Geographical segments

One of the most obvious types of segments that telecommunications firms, especially wireline firms, deal with is the geographical segment (country, state, city, address). Because of the nature of wireline-based services, the customers' residences and where they want the phones installed are some of the most fundamental information about them.

Geographical segmentation has always been one of the most critical components of every telecommunications company's marketing decisions because the cost to install and support the network infrastructure in different geographical areas is the highest expense and presents the largest risk.

Geography tends to play a lesser role in the management of long-distance service (because customers can subscribe to the services regardless of their physical location via dial-in long-distance access numbers). It also has much less significance in the wireless and satellite services markets, because of the nature of wireless service itself and the existence of roaming capabilities.

But because most telcos have been (and still are) organized along geographical lines, and because regulatory agencies still tend to function based on geographical boundaries, geographical segmentation will probably continue to be one of the core structural segment approaches for some time to come.

Product segments

A second and equally obvious segmentation scheme is based on the products the customer buys. Obviously, if a company has several different products and services to offer, there is a very good chance that different people will be the buyers and users of different products at different levels. Ironically, the study of product profitability and product performance analysis is really nothing more than a specialized form of customer segment analysis. Looking at how much money gets spent on cellular phone traffic (for example) is really telling you how much the customers of the cellular phone business are spending. However, because profitability is usually ruled by geographic parameters (very product-specific economies of scale and rules of engineering), we find it convenient and useful to treat the segments created by all of those customers who buy the same product to be a special kind of segment.

Commercial segments

The other major category of structural segments is the basic type of customer that a person or organization falls under. These segments help us separate our different types of customers based on how they themselves do business. We therefore refer to this as segmentation by type of commerce or *commercial segments*. Commercial segmentation is one of the company's main sources of information, especially when it comes to questions of organizational adjustment and segmentation.

Customers who do business with the telco in fundamentally different ways are often better served by different kinds of customer service, sales, and network capabilities. Therefore, commercially based segmentation is one of the first forms of segmentation used by any telco. At a minimum, a telecommunications firm will recognize the difference between consumer (individual private citizen) and organizational customers. Consequently, most telcos have consumer customer organizations and business organizations, each geared to service the needs of these very diverse groups of customers.

Often the organizational customers will then be separated into even more discrete units, which may include the following categories:

- *Large business organizations:* usually referred to as corporate accounts;

- *Government and public services organizations:* often referred to as institutional or governmental or public sector customers;

- *Small and medium businesses (SMB accounts):* a group of customers that falls somewhere in between consumers and large businesses in terms of their needs and buying patterns.

For telcos that are larger, or more specialized, the categorizations can be taken even further. Some telcos go so far as to segment companies by industry, creating delineations between manufacturing, finance, retail, distribution, transportation, and other types of customers.

Why is understanding structural segments important?

You might think that the category of structural segmentation is redundant and unimportant because, after all, the geographical, product, and commercial categories could also be considered categorical or behavioral classification schemes. Why then do we place them into this special and separate category?

There are several reasons. First, these schemes are so basic and so intrinsic to the way that telecommunications firms work that almost without fail every single telecommunications company will work with these approaches as a default. No matter which telco you happen to look at, these segmentation schemes will always be recognized.

Second, these schemes are so intrinsic to the business that they are almost without fail an institutionalized part of the business. This means that organizations are usually built around the management of the different groups of customers who fall within those segments.

Third, because they have become institutionalized, people often overlook the fact that they are only another set of arbitrary segmentation schemes. There is nothing sacred about these segments, and no reason why the marketer should assume that they are not to be included and worked with like any other.

Creating structural segments

For the most part, the marketer will have very little work to do when it comes to creating structural segments. By their very nature, structural segments are intuitively obvious in most cases and therefore need no mathematical or statistical verification.

At times, however, marketers will be called on to recommend the creation of new structural segments (when a major group of customers is not receiving the attention they deserve, for example) or to verify whether continued use of an existing structural segmentation is an appropriate use of resources (such as when one customer group has become too small for a separate organizational structure). In these situations, the marketer will take advantage of the techniques used to derive and substantiate categorical segments to get the job done.

Categorical segments

The structural segments we have discussed are probably the least understood and most often overlooked of the different types of segments that a telco will work with. Conversely, this second group, the categorical segments, is probably the best understood and most utilized. When people think about segmentation, categorical segments are what most often come to mind.

In general, categorical segments are groups of customers who share a common set of characteristics for a given set of criteria. In other words, categorical segments group customers based on the similarities and differences of the people (or organizations) themselves.

Simple (univariate) categories

The easiest categorical segment to understand, and a good place to begin, is the simple type of categorization. *Simple segmentation* occurs when we identify one characteristic about a customer and use it to separate the population of all customers into two or more exclusive groups. To perform categorical analysis, all we need to do are these simple steps:

1. Identify the customer characteristic that we want to use as a segmentation criterion.

2. Figure out the value for that characteristic for all of the customers.

3. Then divide the customers based on their membership in the appropriate group.

Let's use *gender* as our simple criterion and see how this works. We know that all customers will fall into one of two gender categories, male or female. To figure out which customers belong to which group, we assume that our customer information system has that information and all we have to do is find the field in the database that holds the gender variable and use it to "tag" the individual customers. We then divide the customers into their appropriate groups, count how many there are in each category, and report our results.

Simple segmentation reports

The results of segmentation work are most often organized so as to report one of three sets of information, (1) the number of customers within each segment, (2) the amount of revenue or profit associated with each segment, and/or (3) the amount of product utilization associated with each segment. The report itself is done in either a tabular or graphical format. In the case of our gender segmentation, a pie chart to indicate the number of customers in each segment is probably more illustrative than a tabular report. In the pie chart for our example shown in Figure 10.2, we see that we have 2,500,000 female customers and 2,500,000 male customers, or that females represent one-half of our customer base and males represent one-half.

Naming simple segments

In the case of simple segmentation, naming the segments is usually pretty straightforward. We refer to the segmentation scheme by the name of the characteristic used to create it (in our example, this is the gender segmentation). The different segments of the scheme will be given the names associated with the customer's membership in the segment. In our case, the male and female segments.

Statistical basis for segmentation

So far, we have been describing the segmentation process in very simple, nonstatistical terms. However, as we begin to deal with more and more values and characteristics we quickly move beyond the area where simple

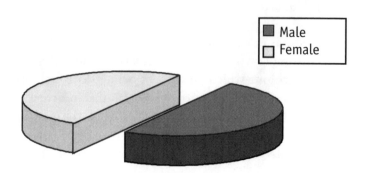

Figure 10.2   Segmentation by gender.

descriptions will help us. The use of basic statistical vocabulary and approaches will be much more useful in the continued manipulation of segmentation information. We will, therefore, look at the process in more mathematical and statistical terms.

The process for creating the simple gender segmentation scheme can easily be restated in statistical terms. Statistically, we could say that we identified a population (the collection of all customers) and a specific variable associated with that population (in this case, the variable is whether they are males or females). We then used this variable to break the population down into two subgroups based on the value of that variable.

This form of segmentation is known as *univariate* or *single-variable* segmentation because only one variable is used to separate the population into smaller groups.

Other forms of univariate segmentation

The gender variable is, of course, one of the simplest of all segmentation schemes because it very neatly divides the population into two exclusive categories. Real univariate analysis often gets more complicated than that.

We can create univariate structures based on customer age, birth date, income level, social status, or any of a number of other discrete values. In these cases, we often end up with more than two categories into which the customers can be placed.

For example, a segmentation by age might yield 100 categories, one for each year of a person's life. A segmentation by the number of years the customer's company has been in business might yield several hundred categories. In all cases, however, the segmentation is univariate only if the segmentation is limited to one variable (or characteristic) that is applied to all customers.

Choosing variables for the segmentation process

The preceding discussion introduces several other interesting points about the process of choosing variables for the creation of segments.

Even when we are dealing with the simplest univariate segmentation case, we find that we have some challenges. Our first problem is that the variable we choose to create a segmentation might yield so many different values that the segmentation approach fails to simplify anything.

For example, a segmentation by age, with 100 age categories, will not be very useful. The usual approach is to identify ranges of values to create the segments. We could divide the ages 1–100 into categories of 1–18, 19–25, 26–55, 56–100. These four categories allow us to learn some interesting things about these customers as a group, while still minimizing the number of segments.

The other problem arises when the chosen variable is not a simple categorical variable to begin with. For example, suppose we want to create a segmentation based on customers' political beliefs. In this case, there is no single variable that tells us what those beliefs might be. We have to try to figure out how we might *derive* the value of the variable, based on other variables that we do have. We might know what magazines a person reads and what clubs the individual is a member of. With this information we could possibly figure out the person's political standing. These types of variables are called *derived variables*.

Techniques for the creation of univariate segments

In general, therefore, we find that there are three types of univariate variables that we can use to create our segments and several different ways to create or discover them:

1. *Explicit:* having values for a specific customer variable (i.e., gender: male or female);

2. *Range based:* categorized based on the value of the variable found within ranges set by the marketer (i.e., age: 1–18, 19–25, 26–55, 56–100);

3. *Derived:* having a value that must be derived based on the values of other (i.e., political standing: conservative, moderate, liberal).

The first—and easiest—way to create a segment is simply to choose a characteristic that you know you have data for (i.e., you have a variable available). This process is usually done by (1) asking the IT department to create a report, (2) using a query tool to pull out the information your-self, or (3) using a spreadsheet to help you wade through the data and create the groupings. This technique works best when you have a good source of data to work with and the variable is explicit.

When the value for the variable must be placed into a range, or derived, the marketers are usually faced with two options. They have to learn a programming language and write the programs that create new explicit variables and fill in the values, or they work with a program-mer who can do it for them.

Minimizing categories to maintain clarity

The main objective of segmentation is to take a large number of very different customers and categorize them into much smaller groups. From a purely mathematical perspective, we can see that, before the segmenta-tion process, there are two clear segmentation schemes that function as the defaults. The first scheme makes no differentiation between custom-ers and there are only two segments, customers and noncustomers. In the second scenario, we have at least as many segments as we have customers since we can view the entire population of customers as each comprising his or her own unique segment.

One-to-one marketing

In fact, there is a school of marketing in vogue today that believes the ulti-mate goal of marketing is to consider the entire population of customers to be a vast assortment of different markets, each to be handled in a differ-ent way. Today's technology actually makes that kind of marketing stance possible for many companies, and it is certainly a goal to consider.

When companies make a full-fledged and serious commitment to the development of integrated marketing systems and customer relationship management software, then this kind of vision becomes a reality.

However, most of the world and most telecommunications firms are still a long way from being in the position to invest in and make use of these kinds of fully integrated systems. And even after they are fully operational, it will be some time before the typical telco is ready to give up on the many benefits that segmentation still has to offer.

11

Complex Segmentation

The more the merrier.
Anonymous

In the previous chapter, we became familiar with the concept of segmentation and saw how some of the simpler schemes are developed, displayed, and managed. In this chapter, we will get into some of the more sophisticated and advanced forms of segmentation and see how the telecommunications marketer includes several different variables in the segmentation efforts.

Complex (multivariate) categories

The easiest kind of segmentation to deal with is the univariate type, because you work with only one categorization at a time. However, univariate segments provide us with only the most elementary kinds of customer differentiation. It is impossible to learn who customers are by looking at only one aspect of their characteristics. People are complex and so too must our segmentation approaches be.

Complex segmentation occurs whenever the marketer combines two or more categorical variables in the same segmentation study. Of course, as you add more and more variables, keeping track becomes much more difficult.

Examples of complex segments

A few examples should help illustrate.

Let's assume that we have segmented our customers by gender and found that women buy more cellular time than men by a ratio of 3:1 (see Figure 11.1).

What more could we learn if we understood what age groups within these two major categories were contributing most to the revenue generated? To find this out, of course, we need to make our segmentation scheme more complicated by including the age variable. For the sake of clarity, we will divide our population into three age groups: 0–21, 22–55, and 56–120. Figure 11.2 shows the revenue distribution after we add this criterion to the gender segments.

By looking at the distribution of revenue across both the gender and age criteria, we obtain a much clearer picture about exactly who the real revenue producers are for our cellular business. As you can see in Figure 11.2, the category of females between the ages of 22 and 55 represents $2 million or exactly one-half of all cellular phone revenues.

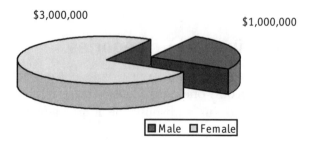

$3,000,000 $1,000,000

Male Female

Figure 11.1 Cellular revenue by gender.

Challenges in the representation of multivariate segments

Our preceding examination of only this most basic kind of multivariate segmentation scheme, a bivariate scheme (*bivariate* means that two variables are used), demonstrates how much more difficult it is to work with multivariate segments. Even in this simple case, we found that we had to change our style of presentation in order to visualize the different pieces of the segments. The pie chart was no longer effective because there were too many pie slices and some were of such similar size that the display would have been meaningless.

Indeed, as you add more and more variables to the multivariate analysis, the challenge of presenting your findings and keeping track of all variables involved becomes increasingly complicated.

Naming multivariate segments

Naming multivariate segments is another area of difficulty. Coming up with names for univariate segments is easy because you can use the name of the variable or the value that the variable holds.

For example, if we were doing a segmentation by gender (using the variable *gender*) with two populations (a group of males and a group of females), then naming it the *gender segment* would be quite descriptive. When we get into the use of multivariate segments, however, this process is no longer so simple. We can start by stringing the names of

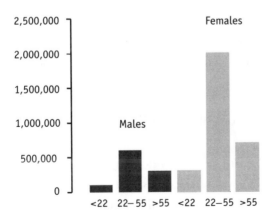

Figure 11.2 Revenue by gender and age.

variables (the gender/age/income segmentation, for example), or we can create acronyms (e.g., GAI for gender, age, income segmentation).

But as we get increasingly sophisticated, combining dozens or even hundreds of variables in the calculations, even that process becomes unwieldy. Ultimately, the marketer is left with the job of creating interesting, catchy, informative names that can clearly communicate what the segmentation is all about. This is how acronyms such as YUPPY, DINC, and others are contrived and where the marketers can get very creative.

Techniques for creating multivariate segments

The world of the marketer gets much more interesting as soon as we create and display information about multivariate types of segments. In the multivariate case, we combine the issues of managing simple univariate segments with the complexities of dealing with many of them at the same time. As we saw in the case of our simple gender versus the gender/age segmentations, the display of the relationships between the segments also gets harder to do. There are, in fact, three basic ways in which we can develop and study the composition of the multivariate segmentations:

1. Recursive reporting techniques (control-break, cross-tab, and OLAP);

2. Classical statistical methods (cluster analysis, factor analysis, CHAID);

3. Neoclassical statistical methods (neural networks, regression trees).

Let's take a closer look at each of these.

Recursive reporting techniques

As soon as there is a need to track and manipulate information about more than just a few variables, the use of mathematical or statistical techniques becomes necessary. In the univariate and simpler multivariate cases, a person can usually keep track of what variables go into the segment and what the different values mean, but as you go from 1 to 2 to 3 to 7 to 10 variables you can quickly become overwhelmed.

The three techniques used by marketers to manage this complexity are the cross-tab and control-break reports and OLAP (on-line analytical processing) systems.

Ancient (reporting-based) segmentation techniques

In the "good old days" of direct marketing, before statistical analysis packages and data mining tools, marketers relied on recursive reports as the sole source of information for the derivation of segments. The marketer would request that the IT department run dozens or hundreds of different variations of the reports, using various combinations of variables. They could then review and compare the information provided by the different reports and choose the optimum mix.

Nowadays, of course, marketers use more advanced techniques for the majority of their segmentation work. However, these reports can still be extremely helpful for reporting and analysis purposes.

Cross-tab and control-break reports are two segmentation reporting techniques that allow us to display information using three, four, or even more dimensions. These reports were developed over the years to help marketers understand what goes into the different segments. The format of this type of report is very simple and straightforward, but it goes a long way toward helping people conceptualize complex relationships.

Control-break reports

Control-break reports (see Table 11.1) show the different relationships between variables by using two techniques:

1. "Nesting" of data underneath indented report header lines, where each indentation marks a different reporting dimension;

2. Tallying totals for each of the subcategories within the report.

As we can see in Table 11.1, we can find out how many customers are within each of the segments by simply scanning through the report header lines until we find the item we want. The values for the larger "segments" are found on the control-break report lines. We can also see by this report that as the number of variables increases our ability to conceptualize

Table 11.1

A Control-Break Report by Gender/Age/Income

Gender	Age	Income	Number of Customers	Sub-total	Total
Male					
	0–21				
		0–40,000	257		
		40,001–75,000	785		
		75,001–250,000	1,256		
Total males 0–21				2,298	
	22–55				
		0–40,000	745		
		40,001–75,000	437		
		75,001–250,000	146		
Total males 22–55				1,328	
	56–120				
		0–40,000	5,578		
		40,001–75,000	4,433		
		75,001–250,000	543		
Total males 56–120				10,554	
Total males all ages					14,180

relationships increases. Also, our ability to go across control breaks is limited. For example, how many total people (male and female) are in the lowest income group? What the report doesn't tell us, we have to add up for ourselves.

Cross-tab reports

Cross-tab reports represent relationships in a slightly different way than do control-break reports (see Table 11.2). In cross-tab reports, the complex relationships between variables are translated into two dimensions, creating a grid box. The titles of the grid rows and columns are the names of combinations of variables, and the intersection of the two shows the value.

Notice how the cross-tab report of Table 11.2 shows the same information as the control-break report, but in a more compact and

Table 11.2
A Cross-Tab Report

Income	Male 0–21	Male 22–55	Male 56–120	Total: Income
0–40,000	257	745	5,578	6,580
40,001–75,000	785	437	4,433	5,655
75,001–250,000	1,256	146	543	1,945
Total: Age/Gender	2,298	1,328	10,554	14,180

informative way. Unfortunately, the cross-tab report, like the control-break report, is only helpful in the analysis of a smaller number of variables. These techniques are extremely useful in helping identify segments when three, four, or even up to seven or eight variables are involved. But any more than that, and the complexity of the reports makes them unusable.

On-line analytical processing systems

Recently, a new type of on-line reporting tool, known as OLAP or on-line analytical processing, has become popular in many marketing organizations. Basically, these products allow the user to view thousands of pre-defined control-break and cross-tab type reports on the computer instantaneously.

The incredible power and flexibility of these systems now allows marketers to do their segmentation studies without having to ask the IT department to run the reports; they can view all the different combinations of variables on the screen. OLAP systems, however, like the other reporting approaches, severely limit the marketer's ability to move beyond the level of looking at more than a handful of variables at a time.

Challenges to the execution of recursive reporting segmentation

On the surface, it would seem that recursive reporting types of segmentation are simple to perform. This approach is by far the easiest and was once, in fact, the only way to do segmentation.

It is also possible for people to commission and execute small, random runs of these kinds of reports at a minimal cost, with very little investment in technology or specialization of any kind. This will remain true, however, only as long as the marketer is not serious about the exploration of segmentation issues. If the marketer only infrequently needs to look at segmentation issues, then clearly, the ad hoc approach to managing the process will be sufficient. If, however, the marketer finds that more and more segmentations studies need to be done to keep up with the increasing demands made by product managers and market conditions, then the ad hoc, on-the-fly approach will leave something to be desired.

As marketers try to run more and more reports, using more and more data, and in more and more forms, the ability of the existing systems and IT personnel to support them will become greatly strained. The more you expand the scope of the reports, the more difficult it is for the IT department to run what you need.

Data management issues

Unless the marketer happens to be an extremely data-focused, statistician-type of person, the world of recursive reporting segmentation seems to be one of simplicity itself. The marketer tells the IT person which variables should be in the report, and the IT person produces it.

Of course, it is not that simple. In fact, most telecommunications marketing groups are finding that identifying the data and making it available is a very big job. Running one or two reports will happen without too much trouble, but as your requirements for precision, accuracy, and variety increase, the amount of work skyrockets for the IT person.

The reality is that the data, the raw material used by IT professionals to produce the reports, is extremely difficult to find and is usually in very poor condition. The databases and files that support these systems are often many years old, ill maintained, and full of erroneous data. The work involved in cleaning up data to make it usable can be substantial.

Data reporting flexibility issues

The IT professional will have a second problem to deal with when you start asking for dozens of versions of these reports. Unfortunately, most

IT organizations are not equipped to deal with the high rate of change that recursive reporting requires. Most standard reporting tools in the IT area are either large, old, inflexible tools that are difficult to work with, or they are small, flexible PC-based products that cannot deal with the high volumes of data required by these types of reports.

Role of the marketing database in recursive reporting

To ease recursive report analysis, most organizations find that building a marketing database is the most effective solution. These databases, based on data warehousing and data mining technology, minimize or remove most of the challenges to recursive reporting. The basic concept behind the development of these systems is to create an environment where (1) the procurement of data sources is fast and easy through construction of data acquisition systems and (2) the development of reports is made instantaneous through deployment of highly flexible, user-based reporting tools (such as OLAP) that allow the marketer to create as many different reports as often as desired.

In future chapters we will take a much closer look at the role played by these marketing databases in the efficient running of a marketing environment.

Classical statistical approaches

Under the category of the classical statistical approaches, we include all of those ways in which the traditional statistician analyzes large groups of individuals and looks for clues to better ways for grouping them. The category of classical statistical approaches includes factor analysis, correlation analysis, and many other types of statistical exploration and discovery. The most useful and popular of these techniques for segmentation purposes, however, is an approach known as *cluster analysis.*

Cluster analysis

Cluster analysis is by far the tool of choice for most marketers from the "classical" school. Clustering is a mathematical technique that examines very large numbers of variables for common shared characteristics and determines how related the members of the population are to each other. In other words, clustering does the same recursive "hunting for relationships" that the marketer uses in the report method of segmentation, but it does the comparisons on a massive, mathematically accurate scale.

When done correctly, clustering finds and highlights the natural groupings of customers based on the variables they share in common. Clustering, however, is *not* artificial intelligence. It is a simple mathematical tool that identifies groups of people that are related in some way. It does not interpret those relationships in any way and it is the job of the statistician or marketer to review the outputs of the segmentation process and interpret what it reveals.

How clustering works

The clustering process is responsible for identifying the most easily recognized segmentation schemes in the world today, for example, the DINC (dual-income, no children) segment.

Discovery of these segment types came about in a simple way. Marketers collected large pools of information about customers: age, location, income, religion, buying patterns, preferences, and so forth. This information was combined with information about the specific buying behavior they were interested in tracking — in the DINC case, the purchase of high-end sport utility vehicles. The clustering algorithm then performed a comparison of all of the different possible combinations of variable values and pinpointed ways in which the purchasers of these vehicles naturally grouped themselves.

What the analysis showed was that a large group of people was related through a large number of variables. They showed up as a "gathering" or "clustering" of points on the display. Closer examination showed that these people all had two things in common:

1. They were part of a couple (not especially married) who both worked.

2. The couples had no children in the home.

This information—the fact that there was a group called DINCs who tended to be good customers—allowed marketers to target their advertising and direct marketing campaigns at precisely those kinds of people. Figure 11.3 shows an example of a clustering output report.

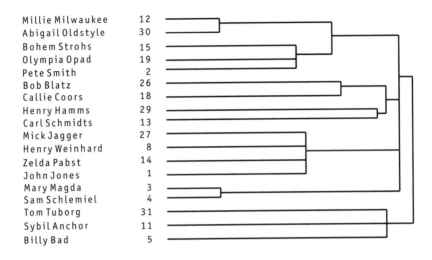

Figure 11.3   A clustering (dendogram) output report.

Dendogram reports show the analyst how closely related different individuals are to each other when all variables are taken into account. Customers having similar characteristics are linked together by lines. The more different the groups are from each other, the further out the lines connecting them go. The report shown in Figure 11.3, a very small subset of a larger report, shows how closely related different groups of individuals are to each other when all variables are taken into account. Certain individuals (like Millie Milwaukee and Abigail Oldstyle) are closely related to each other. This same group is very dissimilar from the other "cluster" of customers at the other end of the scale (Tom Tuborg, Sybil Anchor, and Billy Bad).

Executing classical statistical analysis

Performing classical statistical analyses, such as clustering analyses, is a complicated and sophisticated process. Doing an analysis effectively requires that you do much preparatory work and that you be precise in the way you run the studies. Because of the nature of classical statistics, the quality of the data (its completeness and accuracy) and the volume of data (the number of records you have to work with) must be very high.

In addition, running a cluster analysis or any other form of classical statistics requires that the marketer have access to a statistical analysis software package. Three of the most popular are SAS, SPSS, and IBM's Intelligent Miner.

Role of the marketing database in classical statistical analysis

As with recursive reporting approaches, the organization can run a certain amount of classical statistical analyses without the benefit of a marketing database environment. Certainly, in the early stages of segmentation, it is easier for the marketer to do a quick "down and dirty" analysis without any of the trappings of a full-fledged marketing database.

However, as the sophistication of the team grows, and as the need for more and more clean, accurate, and readily available high volumes of data increases, a marketing database environment becomes critical. What is needed is the ability to do all of the steps required for these studies often and quickly.

Neoclassical statistical approaches

In recent years, the presence of low-cost, high-powered computers and the development of whole new mathematical sciences have turned the field of mathematics and statistics on its ear. Terms such as *chaos theory, neural nets,* and *decision trees* are creeping into mainstream mathematics at an ever increasing pace. These disciplines, referred to collectively as the *neoclassical* statistical approaches, perform the same types of work as the classical statistical approaches, but in newer and better ways.

Neoclassical approaches accomplish what they do by taking advantage of the fact that computers are able to perform trillions of calculations in a relatively short amount of time and at a very low cost. This means that computationally intensive operations that were not feasible in the past can now be executed on a routine basis.

In the case of segmentation, two kinds of products are becoming extremely popular:

1. *Decision trees:* products that organize large volumes of data about customers and build logical "trees" of relationships between the different variables associated with them;

2. *Neural networks:* products that are based on the development of hundreds or thousands of little "programs," each of which functions in a manner similar to the way that a brain cell (neuron) functions. These networks are then allowed to "read" streams of data and to "learn" what the important segmentation groups within the populations are.

Advantages of the neoclassical approaches

The neoclassical approaches are popular with marketers today for many reasons. First, they do not require as much data or as accurate a source of data as the statistical approaches. (In general, the classical approaches require tens or hundreds of thousands, or even millions of "sample records" to produce accurate results.) The neoclassical approaches can function well with less.

Second, the neoclassical approaches are faster than classical methods, both because you do not have to spend as much time preparing the data and because the models do not have to be precisioned and tuned the way classical models need to be. (For example, a typical cluster analysis software product requires that the statistician set dozens of control variables to get a "good" run of the numbers.)

Third, in many cases, you do not need to be a trained statistician to interpret the results of a neoclassical analysis. Neoclassical software products tend to "package" the understanding of the product into the software, while classical products can only be utilized by trained, experienced statisticians.

An example using CHAID

One of the most intuitively pleasing of the neoclassical approaches to segmentation is CHAID, or the chi-square automatic interaction detection approach. The CHAID approach allows the marketer to identify quickly how strong the relationship between different groups of customers is by interactively "testing" the strength of all possible relationships between all members of the population.

Once the system has conducted these millions of calculations, it uses the results of the activity to create a "tree" that graphically shows the nearness and "strength" of the relationships.

This technique is used by marketers to predict how people will behave based on categorical information. Whereas the clustering technique that we talked about earlier tells the marketer *if* individuals are related, it does nothing to help us explain or predict outcomes. CHAID, on the other hand, does include some predictive capabilities.

Using a CHAID tool is simplicity itself. In this example, we use a product called *Answer Tree,* provided by the SPSS Corporation. SPSS manufactures a broad range of sophisticated statistical analysis tools from both the classical and neoclassical categories.

In this case, we have a file containing information about the customers who have subscribed to cellular service. What we would like to know is this: What are the characteristics that are the *most predictive* of this behavior? In other words, which characteristic is the most likely to indicate that a person will respond to an ad? In this case, we know the following:

1. The number of people in the respondent's household;

2. The age of the head of the household;

3. The number of children in the household;

4. The occupation of the head of the household;

5. Whether the household has a bank card or not;

6. The gender of the head of the household;

7. The income of the household.

We want to know which of these characteristics will be the most helpful in future marketing efforts.

The program performs the analysis and outputs the report for review. Figure 11.4 shows a small section of a CHAID output report.

(Our thanks to the SPSS Corporation for the use of their product in the production of these reports.)

At the first level, notice that the number of persons in the household is the single strongest criterion for response. After that, we find that the age and the gender of the head of the household are the next strongest predictors.

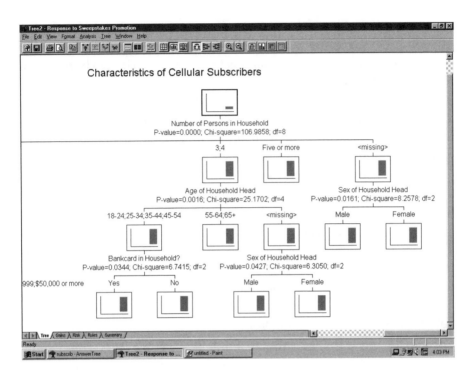

Figure 11.4 CHAID cellular subscriber report.

Compound (multidimensional) categories

Although marketers will find that dealing with structural, univariate, and multivariate models provides them with much useful information, still another level of complexity can be added to the segmentation process. Compound, or multidimensional, categories occur when the marketer finds that it is convenient—or even critical—that an assortment of different variables and categorizations be turned into new superclasses that combine several variables from individual segmentation studies into hybrids.

For example, a marketer may be working with information about customers based on age, gender, and the types of products they buy (a combination of structural and categorical variable types). At some point, he or she may decide that this same collection of variables is being used so often that a special categorization scheme needs to be created to make the data easier to manipulate. This naming of supercategories can be managed in a couple of different ways.

One way is for the marketer to try to come up with unique names to define each grouping of variables (e.g., DINC). Unfortunately, creating unique, cute, and meaningful names for large numbers of different combinations of segments can get to be a pretty onerous and meaningless job. This is when the second approach is used, which involves creating a coding scheme to represent combinations. For example, if we wanted to create a supercategorization scheme for the combination of sex, gender, and product types, we might create a scheme like that shown in Table 11.3. By taking this large number of different variables, and providing a coding scheme to manage them, the job of keeping track of these different combinations becomes much easier.

Super-supercategories and beyond

Of course, there is no reason to think that people will stop at only one level of hybridization in the combination of codified segmentation schemes. In fact, it is very common for people to combine complex

Table 11.3
Supercategories of Segments

Gender	Age	Product	Code
Male	0–25	Wireless	1
Male	26–55	Wireless	2
Male	56–120	Wireless	3
Male	0–25	Long distance	4
Male	26–55	Long distance	5
Male	56–120	Long distance	6
Female	0–25	Wireless	7
Female	26–55	Wireless	8
Female	56–120	Wireless	9

categories into extremely complex "chains" of variables and relationships. The creation of these complex segmentation schemes, though apparently confusing at first, actually serves the very necessary function of helping the marketer to simplify what can quickly become an incredibly complex world of multivariate analysis.

Categorical information: commercial versus consumer

Most of our examples involve segmentation schemes for consumers, but this same approach can be used to analyze and segment businesses and organizations on the commercial side. Businesses have characteristics just as people do. Some are exactly like consumer characteristics, such as age (how long they have been in business), location, and income. Others are characteristics unique to businesses, such as number of employees, type of business, capital assets, and number of phone lines. In both cases, consumer and commercial, the same rules and approaches apply.

Sources of information for complex segmentation

Reporting on and manipulating information about structural segments is usually an easy task because structural segments generally have legacy systems built specifically to keep track of and report on sales activity. Analysis of univariate groups of customers is also fairly simple and straightforward. When it comes time for complex segmentation analysis, however, the challenges get much bigger very quickly.

When marketers decide that it is time for some serious complex analysis, the first challenge is to find the data required and to convert it into a useful format.

There are two sources for the raw material required by the segmentation process. Marketers can go to the company's own internal system, or they can go outside of the firm and purchase lists holding information about customers.

Internal systems

By far the most common place marketers go to for categorical information about customers is their own legacy on-line transaction processing (OLTP) systems. The most common sources of data include the following, in order of popularity:

1. *Marketing database.* If the company has invested in the development of a comprehensive marketing database environment, then there is a good chance that everything they need is stored in this one, convenient location.

2. *Billing systems.* Billing systems often hold key categorical information about customers, including name, address, age, and other variables. The extent of information found there will depend on two factors: (a) how comprehensive the billing system is in capturing the information and (b) how comprehensive the organization is in filling in the fields that hold the information. Unfortunately, in many, many cases, a company's billing system has fields to hold such information, but the users of the billing system will not fill them in.

3. *Customer information system.* Many telecommunications companies have a separate system for keeping track of customer information. For marketers lucky enough to have this kind of system, a wealth of categorical data is usually available.

4. *Other systems.* Different companies may also have other bits of information stored in other operational systems or data marts scattered about the company. The job of figuring out where these sources are and what they hold is a big one that many marketers find frustrating.

Purchased or rented lists

The other place where many marketers will turn for more input for their segmentation exercises is to firms outside of the company from which they purchase or rent consumer or commercial information. There are hundreds of companies in business today that collect information about consumers and businesses, compile that information into large lists, and then sell access to those lists to companies who are trying to enhance their segmentation work. These businesses often do the list brokerage work as a sideline. For example, many magazines keep lists of their subscribers complete with demographics profiles, which they then broker out to people as a side business. Other businesses sell access to their lists to other, noncompeting businesses as an additional source of income. For other companies, the compiling and selling of lists is their only business. In some situations government agency census data or association/society membership lists can be acquired as well. In all cases, the purchased or rented lists are used to add variables to the marketer's own base customer list, in order to enable the marketer to develop deeper and more sophisticated models.

12

Behavioral Segmentation

Actions speak louder than words.
Anonymous

Action is the *only reality*.
Abbie Hoffman

In pursuit of better segmentation

The different kinds of segmentation that we have considered until now (the structural and categorical types) certainly help the marketer and the organization know something about who is buying their products or services and why. These approaches, however, still leave much to be desired when it comes to accurately predicting what people will do in a given situation.

Remember, the objective of segmentation is first and foremost to predict accurately how consumers' behaviors will change based on modifications in stimuli (whether it be an advertising message, a pricing policy, or the firm's presence at a local trade fair). Experience has shown, however, that while structural and categorical segmentation studies can provide some insight and even some casual or inferred predictive capabilities (we will consider this issue at length in our chapter on modeling), they are far from precise in their findings.

When it comes to accuracy and precision, there is a very effective approach that analyzes customers based on the history of their behavior in the past. We refer to this type of segmentation as *behavioral segmentation studies*.

Behavior as the basis for segmentation

As many a wise person has noted throughout history, the best way to figure out what people really think and feel is to watch what they are doing. From the marketing perspective, this is especially true. Although marketers may study the different characteristics of customers—their age, gender, preferences, and attitudes—ultimately, the only thing anyone really cares about is the consumer's actual buying behavior. We only study the other characteristics when they help us understand the real behavior better.

Marketers have learned, over the years, that the absolute best way to predict anyone's buying behavior is to look at his or her past behavior. People are incredibly consistent in their patterns. Consequently, the use of behavior as the means to identify different groups of customers is by far the most powerful and accurate form.

Behavioral segmentation in telecommunications

When you try to figure out how to categorize the behaviors of different customers specifically for the telecommunications industry, you enter a world in which telecommunications marketing is absolutely unique and different from any other form of marketing.

Every company, regardless of industry (banking, retail, services, or telecommunications), basically deals with the same sets of variables and challenges when trying to manage the segmentation of customers. After all, the customers of telcos are the same customers that retailers and banks market to. Our discussions about marketing have also been very generic. The same thinking and approaches apply regardless of which industry you are dealing with.

However, when you enter the area of behavioral segmentation, which, by the way, is the most powerful, predictive, and accurate of the segmentation approaches, you move into a world where there is almost no transfer of expertise from one industry to the next. After all, consumers do not buy clothes the way they buy long-distance carriers. That makes telecommunications customers and their behaviors unique to the industry.

This, then, proves to be the most interesting and challenging of the segmentation areas and the one area where only the telecommunications marketing expert is qualified to participate.

Learning from direct marketing

The direct marketing industry has been in the business of segmentation and prediction of customer behavior for a long period of time. This industry consists of those companies who survive solely on their ability to sell products via mail or the telephone. Catalog department stores such as Sears and Wards, catalog specialty firms like Fingerhut and Land's End, and business suppliers like Viking Office Supply and CDW all make their living by keeping track of consumers and predicting their buying behavior.

Over the years, the direct marketing industry has developed a very precise form of segmenting customers and predicting their buying behaviors. These companies are able to work with incredibly large lists of customer names (often numbering in the tens of millions) and are able to predict how much these people will spend within pennies of the actual expenditures. Indeed, the direct marketing industry depends on this incredibly precise predictability to survive.

Despite the hundreds of statistical and data mining types of software available, despite their ability to work with and manage hundreds of variables, these firms have discovered that all it takes to accurately predict consumer behavior is to use three simple variables: recency, frequency, and monetary.

Recency, frequency, and monetary (RFM)

The terms *recency, frequency, and monetary,* also known as RFM, refer to the three variables that direct marketers have come to depend on as the major predictors of customer behavior. Each of these words represents a different behavioral characteristic of the way people buy. Together, they paint an incredibly accurate picture of the buying behavior of a large group of people. Let's define each term:

- *Recency.* This characteristic tells us how recently someone bought from us. A person who bought from us two weeks ago is very likely to buy again. Someone who bought a year ago is less likely to buy, and a customer of five years ago is highly unlikely to buy.

- *Frequency.* What are the consumers' spending patterns? Do they buy monthly, quarterly, annually? How frequently should they be contacted?

- *Monetary.* How much does the consumer spend each time? Is it less than a hundred dollars, or many hundreds of dollars? A typical business customer might spend thousands or millions.

By keeping track of these three behavior characteristics, the marketer is able to predict who will spend how much in the future.

Methods for display of RFM

The industry-standard method used to display RFM analysis is by means of the old standby cross-tab reports. These reports typically list the monetary ranges across the top, with the recency and frequency values listed along the side. The number of customers in each segment is listed in the appropriate box (Table 12.1).

Based on the information provided by RFM reports, direct marketers are able to predict how much money individuals in each segment will

Table 12.1

An RFM Cross-Tab Report

Frequency (number of purchases)	Recency (last purchase date)	Monetary (Amount Spent)				Total
		$1–$100	$101–$250	$251–$999	>$1,000	
1	0–6 months	265	165	123	57	610
1	6–12 months	174	46	100	33	353
1	1–2 years	32	67	78	32	209
1	2 years	58	29	63	98	248
2	0–6 months	58	47	19	44	168
2	6–12 months	98	67	65	32	262
2	1–2 years	46	190	75	152	463
2	2 years	167	65	75	34	341
3 or more	0–6 months	34	94	51	93	272
3 or more	6–12 months	41	60	12	43	156
3 or more	1–2 years	82	155	26	33	296
3 or more	2 years	34	5	17	12	68
Total		1,089	990	704	663	3,446

spend during the next campaign. The higher the recency, frequency, and monetary, the more likely they are to spend.

Types of telecommunications behaviors

Although the RFM approach may be a great way for direct marketers to track consumer behavior, the model falls short when attempting to provide telecommunications marketers with what they need.

Telecommunications marketers, in fact, are concerned with several different types of consumer behavior.

Subscribe behavior

The term *subscribe behavior* describes the event when a consumer decides to change carriers and subscribe with our firm for a particular service. The subscription behavior of customers in telcos is very similar to the "enter store" behavior of retail consumers. Consumers do not *buy* a retail store, but if they do not decide to enter the store, they will not buy anything. For many lines of business (long distance, wireline, and wireless especially), telcos must get customers to subscribe before any revenue generation will happen.

Because of the similarity between the telco "subscribe" and the retail "enter store" behavior, telcos often borrow concepts from the retail industry to help them accomplish their objectives. Some of the more popular retail approaches to getting customers to come to a store include:

- *Sales.* By holding discount pricing "sales" events, retailers attempt to entice people into their aisles.

- *Promotions.* Retailers often hold special events that invite consumers to stop by and look around.

In the same way, telcos will often feature special, limited time only discounted rate plans and other consumer enticements.

Subscribe behavior is the kind of behavior marketers are most concerned with when dealing with issues of acquisition or churn.

Spending behavior

By far, the behavior of most interest to everyone in the telecommunications company is the customer's *spending behavior*. How much does this customer spend each month on each service? The RFM form of segmentation is an ideal way to capture this view of the customer population. RFM-type reports can show how often the consumer uses the phone and how much the consumer spends when he or she uses it.

Utilization behavior

The term *utilization behavior* describes the different ways of looking at how customers make use of services. Utilization behavior can be described in terms of minutes or seconds of service used, the duration of specific calls, the number of calls, or any of a wide variety of other utilization patterns. Obviously, the amount of service utilized is related to the amount of money that people spend, but the relationship between the two depends on several factors.

Payment/credit behavior

Another aspect of customer behavior that is of critical importance to the telco is how well, how often, and how dependably the customer pays bills. The speed and regularity with which a customer pays has a bearing on how attractive that customer is to the telco. *Payment and credit behaviors* describe how the customers pay or what their overall credit rating is.

Maintenance behavior

A recently added topic to the list of behaviors of telecommunications customers is the *maintenance behavior*. Maintenance behavior tells us how much support over and above the minimum service level the customer requires to stay happy.

Maintenance behavior is especially important for commercial accounts. Large commercial customers often require a significant amount of special customer support that is frequently provided as part of the provision of services at discounted rates. With the combination of low rates and high maintenance costs a commercial customers might look attractive on the surface (because of the high utilization and spending behaviors), but actually may not be so desirable when the total cost is considered.

The customer behavior function, B*f*

Of course, no one customer behavior variable, in and of itself, will provide the information we need. Depending on what we want to know and what we are dealing with, we need to develop a different combination of these factors.

The actual combination of behavior factors and the relationship between them that gives us a useful picture of the customer's behavior is known as the *customer behavior function* (B_f). This function is derived by defining the relationship between our different behavior variables, namely:

- *Sub:* subscribe behavior;

- *Sp:* spending behavior;

- *Util:* utilization behavior;

- *Cr:* credit/payment behavior;

- *M:* maintenance behavior.

The customer behavior function, B_f, describes for us the different variables that apply in a given situation. We can describe the general behavior function equation, therefore, as follows:

$$B_f(\text{Sub, Sp, Util, Cr, M})$$

We can also define the specific functions for different situations by listing the applicable variables. Consider these examples:

1. *Pay phone business.* Customers do not have to subscribe to pay phone business, and it involves no special customer maintenance or credit concerns. (They pay cash.) Therefore, the behavior function for pay phones would be:

$$B_f(\text{payphone}) = B_f(\text{Sp, Util})$$

2. *Wireless flat rate.* In a marketplace where wireless service is offered for a flat fee, the function will involve all of the variables, but the Sp (i.e., the spending variable) will be a constant. We could display this function through the following equation:

$$B_f(\text{flat rate}) = B_f(\text{Sub}, \text{Sp(constant)}, \text{Util}, \text{Cr}, \text{M})$$

We will return to the issue of customer behavior functions in the next chapter when we discuss the modeling process, and again when we look at scoring as part of the campaign management process.

Sources and methods of behavior segmentation

Given how important and valuable behavior information is, knowing where to find it and how to manage it is obviously a key component of any telecommunications strategy. There are actually several "classical" sources of this kind of data. However, each of those sources has its own idiosyncrasies and uses.

The best internal sources for each of the different types of behavior include:

- *Billing system:* subscribe, spend, and payment behavior;

- *Credit system:* payment/credit behavior;

- *Customer service system:* maintenance behavior;

- *Customer information system:* subscribe, spend, payment, maintenance behaviors;

- *Call detail records:* utilization behavior.

Information can also be purchased from outside sources, including these:

- *Payment/credit behavior:* credit agency reports;

- *Subscribe behavior:* external surveys, consumer reports.

Billing and account history analysis

It should come as no surprise to anyone that the telecommunication company's billing system is the number one source for most customer behavior information. The billing system is, after all, the heart and soul of the telco. It is through this system that all customer transaction-related information must pass in order for the customer to be billed and the company to be paid. Because of this, behavioral analysts usually find the best sources of information here.

Finding and using billing information

The first thing that anyone will try to do when faced with the task of analyzing billing systems data is to get the information they need out of the system using its built-in reporting capabilities. Unfortunately, billing systems are not marketing databases, and before too long the analyst usually gets frustrated trying to get the highly limited default reporting systems to do the kind of precision searching and analysis that needs to be done. At some point, this approach will no longer be adequate.

When this happens, the next step will be to extract a copy of all of the information that is of interest to the segmenter and place it into a special marketing database where it can be manipulated at will.

Applications of billing data

Depending on the kind of behavior study you are trying to do, you will need to extract different information from the billing system. In the following sections, we discuss the data needed for each kind of analysis and the approach used to analyze it.

Billing: subscribe behavior

To figure out whether a customer is subscribed or not, all we really need to do is pull a copy of each of the names of existing customers. Anyone who is in the billing system and is active has obviously subscribed. When this information is used in conjunction with other information however, or when the billing system is used to pull information about customers who have stopped subscribing, then a special data column, called "Subscribe," with values of yes or no, can be added to the table. In this way, the analyst can make use of the Subscribe column to assist with further analysis.

Billing: spend behavior

Pulling spending behavior out of the billing system is a little more complicated than getting subscribe information. Spending is a very time-sensitive variable. Therefore, when we attempt to pull spending behavior information, we need to be sure that:

- We understand the time period to which the spending applies.

- We record the spending time period in the database, so that it can be used for future analysis.

- We must be sure that spending information collected from multiple sources (e.g., long distance versus wireless versus wireline) relates to the same time period. (Comparing customer spending on wireless in January with customer spending on long distance in July will not allow us to draw many conclusions about behavior.)

- Spending analysis is based not only on time period, but on product line as well. While the total amount spent with our company is a good spending indicator, spending information by product line is much more informative.

Spending behavior is generally captured in the database in the form of a revenue field. Fields are usually named for the time period and product line being measured. Consider these names: Total_Revenue_January_Wireless and Total_Revenue_March_Long_Distance.

Billing: utilization

Sometimes, utilization information can be found within, or derived from, information in the billing system (depending on the billing system being used). Ultimately, the utilization will be reported as simple monthly totals of service, again, for a product line for a given time period. For example, Total_Minutes_January_Wireless and Total_Minutes_March_Long Distance.

Analyzing and displaying billing system data

Analysis and display of billing systems data is usually handled in a number of ways:

1. *Ad hoc reporting.* The most common way of making billing data available to analysts is probably that of placing the information into a relational database and allowing analysts to view it using ad hoc reporting tools. This approach is the most useful in allowing analysts to begin early exploration into patterns and segments and to verify—at the detail level—assumptions they make about their advanced findings.

2. *Cross-tab reports.* A good snapshot of this information can be rolled up and derived using cross-tab reports.

3. *OLAP.* In some situations, marketers are provided with product line, customer, and time frame dynamic navigational capabilities through the use of an OLAP reporting tool. This system provides the same kind of functionality that the ad hoc and cross-tab reports do, only it is delivered more quickly and with more versatility.

4. *Classical statistical methods.* Analysts often collect a large sample of customer billing data and perform cluster analysis on it, in order to look for specific subscribe, spend, or utilize behaviors.

5. *Neoclassical statistical methods.* In the most advanced cases, neural networks or other forms of neoclassical statistics provide insights into customer behavior patterns.

Time series analysis

Once we begin to collect and tag time-sensitive information, we move into a whole new area of advanced customer study known as *time series analysis.* This form of statistics allows the analyst to look for patterns in the longer range behavior of customers over months or years.

Call detail record analysis

One of the hottest new areas of exploration in the telecommunications behavior area is the analysis of call detail records. No business has better information about the behavior of their customers than telecommunications. Just think about it: Every single phone call that you make, its duration, where it is from, and who it is to are captured by the telco's switching system. Because of this, many telcos rationalize that there must be valuable utilization information stored there.

Finding and using call detail records

Gaining access to call detail records would seem to be a relatively simple and straightforward process. Switches keep track of calls, and then translate, consolidate, and forward that information to the billing system. That is how the telco makes money. When you do an analysis of these records, however, you run into three very large problems.

Volume problems

The biggest single challenge to the analysis of call detail records is the immense volume of data that most telco switching systems generate. Trillions of phone calls are made around the world every year. That means that there are literally trillions of transaction records that you need to capture, store, manipulate, and use. Fortunately, many approaches can address this problem.

Format challenges

The second problem is the fact that the information captured by switches is not usually stored in a very user-friendly and easy-to-manage format. As you can see in Table 12.2, the call detail record holds specific, often codified information about the call that occurs. To make this very granular and specific information useful, the analyst must do some serious data manipulation and computation work.

Table 12.2
Call Detail Record Layout

CALL-FROM-NUM
CALL-TO-NUM
CALL-START-TIME
CALL-DURATION
NUM-OF-LECS

Intervening in the switch-billing system data stream

The key then to capturing and manipulating call detail data begins with getting a good understanding of how the switch data moves from the switch to the billing system, and figuring out which part of the data stream offers the most economic intervention point.

Right now, the existing switch to billing system conversion process exists and runs efficiently. Switch records are pulled, formatted, summarized, and injected into the billing system. If you can take a good look at that process and determine where each step is being performed, you might be able to find some kind of intermediary file where much of the formatting work that you need is already done. You might also find a file with a set of summarized records that offers the required detail without forcing you to manage such huge volumes of numbers.

Managing the cost of call detail record manipulation

The most critical thing to remember about call detail record manipulation is that, because of the large volumes involved, the costs are very high. In our enthusiasm to understand and analyze as much about our customers as possible, we often forget just how large those costs can get. In fact, the value that this kind of detailed analysis can deliver does not always offset the costs associated with it. There are, of course, ways to manage these costs.

Database costs

The single largest expense factor involved in the management of high-volume call detail record systems is the cost of the database that holds the data. Because of the nature of the data processing industry today, people often assume that they need to put these data into a relational database or even an OLAP system in order to do effective analysis.

Nothing could be further from the truth. When it comes to analyzing data, the database that you use to store the data should be selected according to the job you want to get done, not based on what the technicians think is the "right" one to use.

Three types of data storage techniques are available to manage any data:

1. *Flat files.* This is the most common way of storing data, even to this day. Switching systems all store data as flat files, and many billing systems use flat files as their input. Flat files allow for maximum storage on tape and disk at the absolute lowest cost.

2. *Relational databases.* Relational databases are the current data storage medium of choice for most business applications. Relational databases enable users to make ad hoc queries against the data stored there and generate dynamic real-time reports.

   A few important details of relational databases should be considered when we talk about call detail records. First, it costs anywhere from 10 to 50 times as much to store data in a relational database than it does to store it in a flat file. This is because the relational database needs up to 3 to 10 times as much disk space to store the same information as a flat file does and it requires that special hardware, software, and services be employed to build, design, and maintain it.

   Second, as the volumes of data in the database go up, the performance goes down. In fact, many telcos create so many call detail records that no extant database is big enough to store and retrieve them in a decent amount of time.

   Third, many of the tasks that marketers need to perform using call detail records do *not* require the power of a relational database. Statistical analyses and summary reporting are more efficiently done when the less expensive flat files are used.

3. *OLAP systems.* At times, people actually try to put together call detail record systems and store them in OLAP systems. OLAP systems are a special breed of database that allows users to navigate among consolidated views of the detail records stored there. These systems, however, cost much more than relational systems, which already cost much more than flat files. In other words, an OLAP system will store and manage the same volume of data as a flat file system for anywhere from 50 to several hundred times the cost. Unfortunately, there are no OLAP systems available that can manage all call detail records at all levels with any kind of decent performance.

The trick then, when it comes to working with call detail records, is to balance the business value anticipated from the analysis with the cost that managing the data will involve.

Applications of call detail record data

So what do you do when you get the call detail records? Basically, call detail records are analyzed for these reasons:

- Identify the calling patterns or habits of an individual customer.
- Identify "clusters" of customers with the same calling behaviors.
- Identify specific calling behaviors for further analysis.

In all cases, these are the objectives:

- Identify "good," high-value customers. This information is used to help the marketer find more customers with the same kinds of behavior.
- Identify "good," high-value behaviors. This information is used to give the marketers some kind of idea about the kinds of customer behaviors they want to encourage.
- Identify natural segments of customers who can then be sought out and managed as a group.

Ultimately, call detail record analysis provides nothing more or less than another set of variables that can assist the marketer in the segmentation and modeling process.

Analyzing and displaying call detail record data

Call detail records, once they have been formatted, are almost always analyzed at the statistical (classical and neoclassical) level only. A detailed report or an ad hoc report that shows each and every call that a person makes (hundreds per month) would not be useful information for the marketer. Analytical software, such as clustering or neural networks, helps to examine the calling behaviors and to summarize the information in different ways. This kind of analysis can identify groups of customers by characteristics such as the following:

- When they make calls;
- Duration of calls;
- Where calls are made to;
- A myriad of other characteristics.

Credit history, customer service, and other data sources

Although billing and call detail records are the most common sources of behavior data, many other internal and external sources can be referenced as well. These systems, for the most part, face the same kinds of challenges and are used to accomplish the same kinds of objectives as the billing system.

Top telecommunications segments

The last three chapters have covered a variety of different types of segments and an assortment of directions in which the segmentation process can be driven. Certainly, one would think that there are certain segments and certain segmentation schemes that telcos have found to be the most useful and that everyone tends to use. A recent survey, conducted by Tele.Com magazine and Andersen Consulting, asked companies which segmentation criteria they used to help them decide whom to sell bundled services to. The result was a list of the most popular segments, the ones that most companies used to help in the analysis.

Consumer segments

In approaching the consumer marketplace, the top segment criteria, used by more than 40% of the telcos, were all behavior based and included these:

- Internet usage;
- Technology acceptance;
- Monthly phone usage.

The second most popular set of criteria, favored by more than 20% of the telcos, were category/demographics based and included the following:

- Household income;
- Education;
- Age.

The lowest category (less than 20%) involved a mixture of the two:

- Cable usage;
- Occupation;
- Zip code;
- Credit card usage;
- Hobbies.

Commercial segments

On the commercial side of the business, the segments targeted were even less organized or specific. The top category, favored by more than 78% of all telcos, was to segment the customers based on *communications expense* (how much they spend on telecommunications as a firm).

The second set of categories, favored by 30% to 50% of the telcos, involved these characteristics:

- Number of locations;
- Industry;
- Number of employees.

The lower quadrant of categories (less that 30%) included these:

- Revenue growth rate;

- Current revenue;
- Sales expense;
- Zip code.

Data mining versus categorization

The other interesting information that came out of this survey had to do with who used what techniques in managing the segmentation process. They found that different lines of business used different approaches.

Long-distance segmentation preferences

This study found that 50% of the long-distance companies preferred to stick with straight categorization as their approach to segmentation, whereas the other 50% said that they counted on data mining (statistical analysis) to develop their models.

Wireless segmentation preferences

On the other hand, the survey showed that only 10% of the cellular companies were satisfied with the straight categorical approach, whereas 90% saw the use of data mining as a critical competitive advantage. No doubt, these differences are due, in no small part, to the very different nature of the markets in which these companies compete.

How does segmentation fit into the process?

Segmentation is obviously a large, complicated, and powerful process. But when you look at our model of how the marketing process works in a telecommunications firm, you see no "box" that holds the segmentation process. That is because segmentation plays a key role in every one of the processes.

Corporate direction setting

At the corporate level, segmentation plays many different roles. Organizationally, the information provided by marketing about who the different groups of customers are helps define how the company itself gets organized. For example, a study might uncover the fact that the company has a large number of business customers who live in urban communities and who are not being well served by either the large corporate or consumer customer service organizations. Further study might reveal that this is a growing market segment with a lot of unrealized revenue potential.

On the basis of this information, the company might decide to create a separate urban business support unit that can provide better service to this group of customers and capitalize on the revenue being missed. (In other words, segmentation may lead to the creation of additional "structural" segments.)

Typical organizational effects of segmentation include the creation of separate sales organizations, customer service areas, marketing departments, or even completely new business units.

Along with organizational insight, segmentation studies also provide to upper management and the executives in charge of the different operational units valuable information about how well their units are performing with different types of customers. In fact, almost no communication between marketing and the rest of the business occurs without being shaped within the context of some kind of segment-based orientation.

Interface with operational units

Segmentation plays a key role in the management of the relationship between the operational units of the telco and the marketing unit. Segmentation studies provide operational managers with detailed and focused analyses that allow them to see how well they are performing their duties.

Segmentation is also the vehicle through which operational units communicate objectives and goals to the marketing unit. Operational managers will detail to marketing exactly what kinds of changes they want to see in each of the different segments with which they deal.

Prioritization and goal setting

No aspect of the prioritization and goal-setting process could take place if the marketer did not have some kind of segment-based foundation on which to base decisions. Segmentation studies provide the marketer with a clear picture of how well different groups of customers are performing based on revenue, profit, and utilization criteria. This information, in turn, provides the marketer with the initial insights necessary to make sound decisions.

Modeling

By the time we get to the modeling process, segmentation has become more than fundamental. There are three bases for models: (1) segments, (2) product line objectives, and (3) media and messages. But the whole process of modeling starts with the selection of the appropriate segmentation on which the rest of the analysis is based.

Campaign planning

After an appropriate model is selected, marketers use their segmentation studies to form the starting point for the selection of the individuals to whom the marketing message is to be sent.

Campaign execution

The campaign that is actually run will be to send messages to the specified segments over a certain period of time.

Feedback and analysis

Finally, segmentation allows the marketer to analyze exactly how effective the marketing efforts were in impacting the behaviors of different segments.

The segmentation life cycle

Once we understand the critical and dynamic role that segmentation plays, it becomes obvious that any attempts to run a marketing process based on the existence of a single segmentation scheme would be foolhardy and wasteful. An optimum marketing environment, then, is one in which segments are constantly developed, tested, utilized, measured for effectiveness, and then redeveloped. Constant, dynamic, responsive segmentation is the key to running an optimum marketing environment.

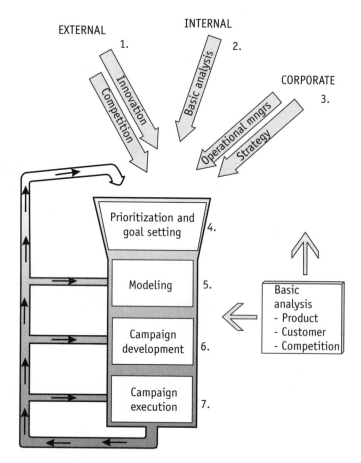

Figure 12.1   Tracking the life cycle of segmentation.

Tracking the life of a segmentation scheme

We can make use of Figure 12.1 to help us understand what the optimum (short but highly effective) life cycle of a segmentation scheme might be.

Exploratory segmentation

Marketers, in working with operational managers and corporate strategists, discuss and report on different ways of segmenting the customer base according to the issues they are currently facing and the objectives they want to meet.

Choosing candidate segments

During the prioritization and goal-setting process, the best of the "exploratory" models is chosen for continued exploration in the modeling phase.

Building of models

During the modeling phase, the best candidate segments are used as the starting point for the development of different campaign models. The objectives of the operational managers (wallet-share enhancement, acquisition, and retention) for different product lines are translated into financial and predictive models. The best segment-based models will become the basis for campaign development. (At this time, many segmentation approaches will be heavily modified and enhanced as the analyst combines more and more financial and predictive features into the model. This will result in the creation of yet another set of segmentations.)

Scoring of models

During the campaign execution phase, the segmentations are enhanced even further, as the likelihood to respond to the campaign score is added to the list of variables associated with the customer. This variable creates yet another segment to which the customer belongs.

Feedback and resegmentation

After the campaign is run, the sources of behavioral information will again be polled, and a fresh set of data pulled into the system. This new information will then show us exactly what impacts the campaigns have had on which segments. Further investigation will help marketers determine which segmentation schemes were the most effective so that they can modify them again to get even better results the next time around.

13

Scores and Functions

In the final analysis, the score is the only thing anyone looks at.

Anonymous

You can't win if you don't score.

Sports truism

In our investigation of the different analytics involved in the support of marketing efforts we have thus far concentrated on approaches that are basically categorical. These approaches define a set number of categories and assist in determining which categories different people belong to.

Categorization is only one of the many ways to organizing information about people. *Scoring* is the second most common analytical approach used by marketers to gain a different perspective on the people they are addressing.

Definition of scoring

The dictionary defines a score as "a number that expresses an accomplishment (as in a game or a test) or excellence (as in quality) either absolutely in points gained, or by comparison to a standard." In other words, a score is:

1. A number

2. Assigned to a person or group of people

3. That helps you assess them

4. In comparison to each other or

5. In comparison to some standard that you set.

A score then, in the marketing context, is a number assigned to an individual or market segment that helps the marketer assess that individual or segment according to whatever criteria the score has been set up to measure.

Different types of scores

There are several types of scores, each of which provides for a different kind of valuation:

- Ranking scores;

- Probability (propensity) scores;

- Desirability (weighting) scores.

Ranking scores (ordinality)

Ranking scores are one of the most common types of scores. Marketers evaluate customers for some factor or factors of interest and then rank the population according to their desirability. (This approach is also called *ordinal ranking*. Ordinal numbers specify the relative position, i.e., first, second, third, and so on.)

The most common application of ranking scores by marketers is scoring lists. Customers are evaluated for how likely they are to buy, and then ranked from the most likely to buy (number 1) to the least likely (number *n*). This ranking allows the marketer to see exactly how the customers spread out around the variables of interest for buying behavior and to make assessments regarding how campaigns might be changed or targets adjusted.

Probability scores

Another common application of scores to a population of people has to do with probabilities and the prediction of people's behaviors. In this situation, marketers apply behavior analysis and forecasting techniques to develop a number that represents the likelihood that an individual will exhibit a certain behavior.

Probability scores are used to evaluate the propensity of customers to buy (in the case of list scoring) and are an incredibly useful technique for the prediction of churn, credit risk, and fraud.

Desirability scores: weighting and customer functions

Another way to utilize scores is as a numerical weighting mechanism for "goodness of fit" against a particular criteria.

For example, as the models become more and more complicated, the viability of simple segmentation approaches breaks down. Or consider this example: You decide to segment business customers by number of employees, line of business, location, and annual revenue, where different combinations of values identify companies as more or less attractive. You will eventually come to the point where you need to codify the segmentation process. By using numerical values to rate each variable, and using a combination of those numbers to create a single score (weighted value) for the company, you can generate a "score" that represents the weighted value of each one. The customer value function, discussed in the next section, is one of the most common applications of this type of score.

Armed with this basic understanding of what scoring is and the different types of scores that can be generated, we are ready to look at the three most common forms of scoring that a marketer is likely to use. The examples that we will look at represent the most common uses of scoring as well as a baseline of applicability. Most other scoring applications are some variation on these three themes.

Customer value functions

The most universally applicable scoring technique is known as a *customer value function*. The customer value function is a score assigned to an individual (or company) that communicates to the marketer, through one number, exactly how desirable or valuable a particular customer is.

Customer value functions have been used frequently in recent years in relation to another function, called the *lifetime value function,* which we will consider momentarily.

Why do we need a customer value function?

When a marketer develops a model to support a certain marketing program, one of the first challenges is to figure out exactly how valuable a particular customer is to the company. The idea of a customer value function may seem simplistic to someone new to marketing. After all, how important could such a calculation be and how difficult could it be to compute? The reality is that the development of this type of score is absolutely essential to the performance of marketing analytics, and that it can be extremely difficult to derive.

But why is the development of such a number so important? The main reasons are focus, simplification, and comparison.

Focus and simplification

Marketing programs are developed to operate with certain groups of customers. In fact, one of the main pieces in the development of any model is determining which segment is best approached to accomplish a certain objective.

Marketers will use many criteria to contribute to that determination. Demographics, behavior, forecasting, and the other ways of looking at the customer are all important and all are taken into account. Most important, however, is for the marketer to develop an understanding of just how critical each of those factors is in relation to each other, against the backdrop of the organization's larger objectives.

The most common way for marketers to handle this kind of ultimate "gestalt" is to include all of those variables in the segmentation process and to use their own judgment. This technique, however, has its down side because different marketers will judge different factors to be important. So it is possible—no, in fact, it is likely—that each model developed will include different factors in the calculation. While the individuality and creativity that this allows may certainly have some merit, this inconsistency between models can have some serious overall consequences. First, some marketers may leave some important criteria out of the calculation. Second, inconsistent development of this aspect of the models means that you will be unable to effectively compare one model to the other.

What is needed, therefore, is a single variable value that consistently identifies the value of a customer to the firm in absolute terms.

Lies, damned lies, and statistics

Looking at the entire process of marketing as we have described it so far, it should become obvious that marketing and its associated analytics are extremely complicated and subjective. Mark Twain once said: "There are three kinds of lies—lies, damned lies, and [the worst kind of lie] statistics."

Comparing one model and campaign to another is a major problem for management. Using this variable will not in any way preclude the use of other modeling perspectives; it will simply provide marketers and managers with a baseline, consistent metric that can be used across all models.

Remember, each marketing program can have a different sponsor, a different team preparing it, and a different objective to meet. Each program will make use of different segments, different media, and different assumptions about what the campaign is to accomplish.

Calculating the customer value function

What kinds of information should be built into this customer value function? And how should it be calculated? Each organization needs to derive the function based on its own criteria. However, basic telecommunications common sense dictates that, as a minimum, the calculation should be based on information found in the customer's basic behavior function.

Remember that we introduced the concept of the behavior function in Chapter 11. We proposed this function as a way to standardize and symbolize how we look at the different types of behaviors that a customer has, so that we can be sure to include those behaviors in the development of our models. Now we want to include that same behavior function in the customer value function.

You may recall that the customer behavior function B_f establishes a relationship between the customers:

- *Sub:* subscribe behavior;

- *Sp:* spending behavior;

- *Util:* utilization behavior;

- *Cr:* credit and payment behavior;

- *M:* maintenance behavior.

When we compute a customer value function, we will want to be sure that these same factors are included. A list of variables and calculations to be included in our customer value function, therefore, would include those discussed in the following subsections.

Number of product lines

The more of our products that a customer buys, the more identified with us he or she is and the more loyal. Therefore, how many different product lines the customer uses is a variable of critical importance.

Weighted value for each product line

We may also want to include a number that gives different "weights" to different product lines. For example, if we determine that long-distance business is highly profitable, but that pager business is not, we will want to appraise long-distance customers as being more valuable than pager customers.

Utilization for each product line
How much of a product a customer uses also affects the overall value represented.

Profitability for each product line
In addition to utilization information (and sometimes instead of it), we need to look at profitability. Customers who use profitable product lines are more valuable to us than others.

Recency, frequency, and monetary
Knowing how long someone has been a customer, how much he or she spends, and how often are also key pieces of information to customer value calculation. The direct marketer's RFM calculations are instrumental in this regard.

Credit and maintenance behavior
In addition to capturing the "positive" aspects of the customer's behavior, we also want to be sure to include any "negative" factors such as credit risk, bad payment history, or poor customer service track.

Creating the customer value function
To determine the customer value function, the marketer needs several things:

- A database or file holding a master list of all customer names;

- A blank field on that database or file that will hold the customer value function;

- Access to all of the other supporting information necessary to calculate the value (usually this will include behavior information, such as spending and utilization, and characteristics information).

Provided with this environment, the marketer can pull together all of the information about the different aspects of customer behavior, history, and characteristics that have been determined to be important, and then create a formula that relates them all. The marketer turns that entire complex mess into a simple, value-based number.

Using the customer value function

Once the function has been created, and the customer database populated with the values, the marketer is ready to make use of it. The purpose of the customer value function is to provide the marketer with an easy-to-use reference that incorporates the relative desirability of the customer into the overall marketing decision-making process. Basically, this number can now be utilized as a shortcut way of appraising the value of different customers across all segmentation and modeling schemes.

Lifetime value functions

Closely related to the customer value function is a similar value referred to as a *lifetime value function*. Whereas the customer value function is a score that identifies the relative worth of one customer compared to all other customers, the lifetime value function is a number that estimates how much revenue a customer represents over his or her lifetime.

Calculating the lifetime value of a customer

To calculate the lifetime value of a customer, the marketer goes through the following steps:

1. Determine the age of the customer.

2. Identify current spending patterns and levels.

3. Project the customer's future utilization and spending levels based on information about how others have spent as they moved through the same age groups.

4. Use that number to determine how much money should be spent to try to keep that customer.

Of course, the calculation of a lifetime value is incredibly subjective and intangible, but many marketers have found the metric useful for appraising many kinds of marketing situations.

Lifetime value calculations are particularly useful with issues of churn and customer retention strategies.

List scoring

While the calculation of customer value functions and lifetime value functions is applicable to all aspects of marketing, this next type of scoring has a very limited applicability. List scoring is a process unique to the direct marketing industry. It is used by the individuals who assemble and run direct marketing campaigns themselves.

The direct marketing process

You may recall from our extensive coverage of the direct marketing process in Chapter 5 that we talked about the scoring process. We said that the basic steps for developing a campaign are as follows:

1. Get a list.
2. Prepare a message.
3. Send the message.
4. Fulfill the orders.

We also said, however, that the real art and science of direct marketing—and the key to efficiency—is to pare down the list of people messages are mailed to, making sure that the ad is sent only to those people who are the most likely to buy.

This process of calibrating exactly who the most likely people to buy are is known in direct marketing as *list scoring*.

List scoring and the customer value function

In many ways, the process of scoring a list and the development of a customer value function are simply two ways of doing the same thing. The objective of both is to provide the marketer with a ranking of just how valuable the customer is. In the case of list scoring, however, we do not care about the customer's overall value, we just want to know how likely he or she is to respond to this particular offer or message.

The list scoring process in action

A critical point about the list scoring process that we need to understand is that the entire process is dedicated to one thing only. That is the optimization of the economic impact of the process of marketing to the people on one particular list.

Although the majority of the other marketing disciplines we have talked about have dealt with more generalized issues, running a list scoring process is very focused. Therefore, to understand this process, you need to know the thinking and setup that came before it and the actions that are taken after it.

Objective of list scoring

The objective of the list scoring process is simple and straightforward. The marketer has chosen a list of names to whom to market. The next step is to perform some analysis, which will allow statistical analysis techniques to rank each person on the list by his or her likelihood to respond favorably to the approach. In other words, list scoring is about getting more precise in your estimation of whether it is worthwhile to send the message to this person.

Economics of direct marketing

Direct marketers would like to see a specific score associated with each name because, to them, the only variable that can be controlled is the number of messages sent. While the broadcast marketer has to accept that everyone will receive the message at the same cost, the direct marketer knows that millions of dollars of marketing expense can be saved if the estimations of customer response can be made more precise.

Scoring a list

First, the marketer has to choose the list. For some telecommunications companies, the selection process is simple—the marketer uses the company's own list. For others, lists may be purchased or leased. In all cases, the marketer will choose the list that most likely offers good candidates.

To score the list, the marketer goes through the following process:

1. Identify customers who have already responded to the offer being made. (For example, if the marketer is trying to sell pager services to long-distance customers, he or she will choose a list of current pager customers.)

2. Analyze that list for segmentation characteristics of interest, including behavioral- and characteristics-based segmentation criteria. Collate a list of *predictors,* that is, variables or characteristics that seem to "flag" the customer as a likely candidate for pager purchase.

   - For simple models, the predictors will be simple "yes" or "no" indicators, stating that if the customer has these characteristics then that person is likely to buy.

   - In more advanced cases, such as models developed with advanced regression, forecasting, or neural network approaches, the system will develop the probability factors based on different combinations of variables. In other words, the statistical analysis software will determine exactly how likely any individual is to buy, based on the patterns that current customer represents.

3. Apply these models of projected customer behavior to the list. The software examines the characteristics of the prospects and assigns a "likelihood to buy" score to each.

4. Check the scores associated with each customer and determine which ones to market to.

Gains charts and "lift"

Of course, presenting a marketer with a list of one million names, each with a different statistical score, is not a very functional approach to doing marketing. After the scores have been calculated, the marketer needs a way to appraise how useful the scoring process has been and to determine whether to (1) proceed with the marketing, (2) try to get a better list, or (3) generate better scores.

Marketers use a tool known as a *gains chart* that shows them exactly how effective a scoring process has been and allows them to make their go/no go decisions.

Gains chart: random response rate

To understand the gains chart, you must first understand the random response rate model (see Figure 13.1). The gains chart basically shows the marketer what the response rate for customers will be for a given campaign. The numbers across the body indicate the percentage of people mailed to. The numbers along the left side indicate the number of people that will respond positively. What this random results line shows us is that, for any campaign that is executed, where no scoring is done, the overall percentage of responses that a campaign will receive will correspond perfectly to the number sent.

In other words, in a random mailing case, if we mail 10% of the pieces out, 10% of the people who are going to buy will respond. When we mail 20%, 20% of the buyers will respond, and so forth.

Gains chart: postscoring

When we "score" a list, we try to figure out who the most likely buyers are, and then send the mailings to them first. If we ranked the list and found that there were 1,000 people who were 100% guaranteed to buy, then we might decide to send the ad campaign only to them and forget the rest.

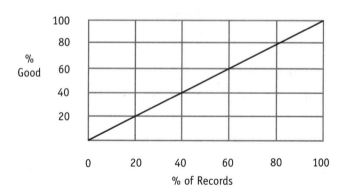

Figure 13.1 Gains chart: random response rate.

Of course, almost no scoring technique can predict 100% likelihood to buy (although there are some cases where it has happened). However, it does make sense for the marketer to concentrate on mailing only to the most likely buyers, and to defer the sending of ads to people who are highly unlikely to buy.

The gains chart of Figure 13.2 shows a normal response curve for a scoring exercise. What you can see in this second chart is a curved line over the top of the base "random results" line. This second line represents the "lift" achieved. It shows the much improved response rate that mailing according to the scores will deliver.

The chart in Figure 13.2 shows us that approximately 60% of the people who are going to buy from this list will be reachable after reaching out to only 20% of the list. Another 20% of the buyers will be reached when the next 20% is contacted. In fact, marketing to the last 60% of the list will only gain you the last 20% of the customers.

What the gains chart shows us is that we should market to only a small percentage of the entire list to get big results.

Cutoff lists

It would be nice if we could get even more specific in our calculations and identify at exactly what point the sending of a direct marketing message is probably a waste of money. In other words, out of my list of 100%, what percentage is the cutoff point for a return on my investment?

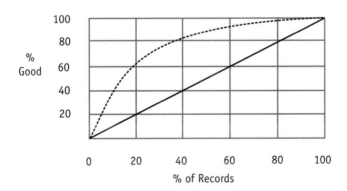

Figure 13.2 Gains chart: excellent results.

To calculate such a point, marketers have developed what are called *cutoff lists*. Cutoff lists take the information generated during the scoring process and project the results out even further. Instead of simply estimating whether the customer will respond or not (a simple yes/no case), the model is enhanced to calculate how much the customer can be expected to spend. It is then possible to compute precisely how many customers should be marketed to. Cutoff lists hold the results of these calculations and identify the cutoff point.

Credit and fraud scoring

The last type of scoring for us to consider is credit and fraud scoring. Exactly the same kinds of techniques are used here to identify who is the most and least likely to exhibit the targeted behavior.

Remember, ultimately, segmentation and scoring are about using statistics to help predict customer behavior. Whether we are trying to predict subscribe behavior, utilization behavior, fraudulent behavior, or poor credit behavior, the same analytical and procedural approaches will apply.

Part 4

The Marketing Process

14

Prioritization and Goal Setting

The way a team plays together is what determines success.
Babe Ruth

Individual commitment to the group effort, that's what makes a team work, a company work, a civilization work.
Vince Lombardi

A closer look at the marketing process

The way the marketing process is organized and executed, especially as it is practiced in the telecommunications industry, can be very confusing. In earlier chapters, we described that process as consisting of several major steps:

1. Prioritization and goal setting;

2. Modeling;

3. Campaign development;

4. Campaign execution;

5. Feedback.

We did not discuss, however, exactly how these different steps occur and how they relate to each other.

Marketing: a continuous melee of a process

Most organizations have some sort of official marketing budget management process, which follows the same planning cycle as any other part of the business. In reality, however, the marketing process does not follow any kind of timetable. Marketing and the evaluation of inputs from the competition, customers, corporate strategy, and operational managers are continuous procedures. The prioritization and goal-setting process, therefore, has no beginning, middle, or end to speak of. It is simply going on all the time.

Because the prioritization and goal-setting decisions are constantly being reevaluated based on a continuous stream of inputs, so too is the rest of the marketing process. The modeling process especially is subject to this demand. There is no moment in the life of marketers when they are not in the process of evaluating one form or another of investigation or segmentation of campaign information in order to help formulate plans or initiatives.

This confusion is made even worse by the fact that most telecommunications firms have numerous people responsible for so many different aspects of the marketing process that no one is ever really sure who is accountable for what and when.

The marketing "team" and "project" concepts

To appreciate the perplexity you need to realize that marketing activities are not executed by a single *marketing organization* that follows a rigidly defined *marketing process*. In most telecommunications firms, loosely defined teams of people perform most marketing work. Typically, these teams are made up of people where some of them work for the same marketing group, some for other marketing organizations, some for outside consulting firms, and some are not part of marketing at all. In all cases, the team of individuals is assembled on a case-by-case basis to address the different projects that come up.

Who defines marketing projects?

Not only is marketing work performed by a hodgepodge collection of individuals, but the very projects that these teams work on are defined by an equally eclectic collection of project sponsors. Marketing projects can be sponsored by:

- *Upper management.* These projects are typically the large-scale, broad sweeping efforts that eventually result in a major campaign initiative.

- *Operational managers.* Often, the managers of various product lines, geographical areas, sales channels, or customer service groups will fund and sponsor initiatives specific to their needs. (These projects may or may not include staff from the marketing department.)

- *Different marketing organizations.* Many telecommunications firms have several marketing groups. Each of these organizations may also sponsor its own projects.

Marketing mayhem

It is no surprise, therefore, that the pursuit of marketing initiatives in a telecommunications firm can be very chaotic. With so many different people pursuing so many objectives and with no cultural or organizational structure in place in order to direct and coordinate these activities, marketing projects tend to be disorganized and confusing.

Marketing project: a definition

A marketing project is a loosely defined set of activities, performed by a marketing team that has been commissioned to develop a marketing plan that will best meet the marketing objectives established by the project's sponsors.

Of course, as we said earlier, the actual focus, formulation, and execution of different marketing projects will vary tremendously from one organization to the next. Sometimes, the people on the team will know the objective, the method, the product, the time frame, and the segment they are targeting before they start. On other occasions, they will be asked to get started with only partial information.

For example, the wireless division may ask the marketing department to figure out how to increase revenue by 10% in the next year. Note that this information provides the product (wireless), the objective (10% revenue increase), and the time frame (1 year), but no details about the method to be used or the segment to be targeted.

In another case, the marketing department may be asked to help design a campaign to increase the number of long-distance customers by 25% in the commercial market in the next 3 months. In this case, they are provided with a product (long distance), a method (acquisition), a major segment category (commercial), and a time frame, but nothing else.

Objectives of marketing projects

As mentioned, then, the marketing project is carried out by a loosely assembled team of people who have been asked to investigate the best means possible to accomplish the objectives set by the sponsors of the project. The team will have a very specific set of parameters to work with (time frame, product, segment, and so on) or a very loose set. In any case, the steps are as follows:

1. Identify the alternative means available to meet the objectives identified to the team, based on the constraints provided. (Develop alternative marketing plans.)

2. Make recommendations for the best way to meet those objectives. (Recommend a plan.)

3. When a clear best solution is offered and where management deems it appropriate, formulate the draft for a specific campaign that will turn that plan into a reality. (Execute a plan.)

Ultimately, the objective of the marketing project team is to provide management with enough information to make good decisions about what kinds of campaigns should be developed and why.

Participants in the marketing process

As we continue to describe this "cosmology" for the nature of the world of marketing, we encounter many new terms and concepts. We will describe a few of the critical terms while we review this organizational view of the marketing process.

Marketing project sponsor

A marketing project sponsor is any member of the firm's management team (upper management, marketing management, or operational management) who identifies a specific marketing objective and who commissions, through the allocation of time, money, and other resources, a team of people to investigate and execute it.

The marketing project sponsor is the person who sets the process into motion, who determines what the objectives are, their importance, and how the marketing team should proceed.

Marketing team

A marketing team is a group of individuals who have been commissioned by the project sponsor to assist in the development of marketing plans and their execution.

Marketing teams usually include marketing experts (who understand advertising, marketing strategies, and so on), marketing analysts (who understand how to run marketing analytics such as segmentation and modeling scoring), database marketing specialists (when a direct marketing campaign is part of the plan), line-of-business experts (who represent the different lines of business involved in the plan), computer systems specialists (who understand how to harness the power of computer systems to assist the team in the pursuit of its objectives), and any other individuals who provide value in the development of the plan.

Campaign objectives

Because the whole point of a marketing plan is to accomplish certain marketing objectives, we need a precise definition for that term.

A marketing objective is a statement by the marketing project sponsor that identifies for the marketing team exactly what effect the marketing activities being considered are to have on the company's market, profit, or revenue position. Marketing objectives, therefore, always call for the increase, maintenance, or controlled decrease of the company's performance in the areas of revenue, profit, or market share. Marketing objectives also usually specify the time frame within which the change is supposed to occur, the product line or lines that are to be affected, and the segments of customers to be operated on (i.e., geography, demographics).

Marketing objectives, can be very broad in scope and general:

> For example, an executive may say, "We need to increase revenue by 10% for our organization overall. I don't care how you do it, just get that revenue up."

Or they can be very focused and specific:

> For example, the manager of the consumer long-distance business may say, "I want to see a 10% in market share for our Midwestern sales office."

In either case, a specific impact on market share, revenue, or profit must be stated.

In summary:

> A marketing objective is a statement, made by the project's sponsor, that identifies the nature of the change in business that the sponsor is looking for (increase, maintenance, controlled decrease), the business metric that is to be affected (revenue, profit, or market share), and the

constraints the team is expected to work with in the pursuit of those objectives (product lines, segments, and time frame).

Marketing WAR: wallet share, acquisition, and retention

As we discussed in Chapter 2, there are three basic approaches when it comes to working with customers and prospects. The team will be able to meet objectives when they figure out the best way to accomplish these tasks:

1. Get existing customers to spend more money, thus improving profits/revenue (wallet share).

2. Gain more preferred customers, which increases the marketing share (acquisition).

3. Keep good customers from leaving (retention).

The job of the marketer can be distilled down to these three core approaches.

Marketing plan

Given these clarifications, we are ready to formulate a definition for the marketing plan itself.

> A marketing plan is information about how the company can best meet the marketing objectives provided by the project sponsor.

A good marketing plan identifies the five M's of campaign development (merchandise, media, market, message, margin) that will best deliver the desired results.

Marketing projects and the prioritization and goal-setting process

The prioritization and goal-setting part of the marketing process is the most confounding to keep track of in any kind of logical order. People are continuously starting projects, ending projects, and changing directions. This process represents the principal point at which the external forces working on the telco (the motive and spoiler forces) are brought into focus and acted on. Creating projects is how the telco takes action in response to those external forces.

The sequence for executing projects and the formality used vary by organization and situation but, in general, prioritization and goal setting follow these steps.

Identification of sponsors

First and foremost, the process of marketing springs from the hearts and minds of individual sponsors. Although it is true that market pressures and cold hard numbers drive the process, nothing will happen unless some individual takes responsibility for doing something about it.

As detailed earlier, the three major sources of sponsors are upper management, operational management, and marketing management. Regardless of the origin, the sponsor must declare an interest in the investigation and pursuit of some specific area; otherwise nothing will be done.

Creation (funding) of projects

The next step is for the sponsor to figure out how to fund the activity. Funding may be the allocation of budget dollars or simply the assignment of resources to work on the task. In either case, a certain amount of resources is dedicated to developing the plan.

It is here in the area of funding that the real problems and confusion within the marketing process arise. Sponsors' wishes are one thing, but the dedication of people, resources, and funds is another, and many projects are started and then abandoned simply because the sponsor was unwilling or unable to continue to fund the activity.

Assembly of teams

Once the levels of effort and funding have been determined, the project team is assembled. At this point in the marketing process, the team usually consists of only those individuals tasked to investigate the project plan at a higher and more strategic level. As the project migrates from prioritization to modeling to campaign development to campaign execution, more and more people and resources are added to the team.

Identification of objectives

The first and most difficult job of the team during the prioritization process is to determine the specific objectives for the project. Clear and precise objectives enable the team to decide whether to proceed and what the best plan of action will be.

It is during this part of the process that the actual prioritization and goal setting occur. Goal setting is accomplished through the specification of clear objectives for the project. Prioritization happens when the team determines how, when, and why to proceed.

Identification of constraints

Finally, the constraints (financial, time, product, segment, and so on) that the team must take into account when developing a plan must be identified.

From a functional perspective, prioritization and goal-setting could also be called the team formation and objective setting step.

Initial definition of the campaign components

Having defined the objectives for the campaign, the marketing team is then ready to begin assembling the pieces of that campaign.

It is the nature of campaigns to be incredibly fluid and flexible and, therefore, we cannot specifically declare how well defined each of these components will be through the lifetime of a typical campaign. Sometimes the organization formally defines each component each time. At other times, people will not even think about or mention certain elements. However, the components are there in all cases. When people do not explicitly define components, then they are simply accepting the definitions that conditions place on them.

As stated earlier, the five M's of a campaign are (1) merchandise, (2) media, (3) market, (4) message, and (5) margin. We briefly consider each here.

Merchandise (What are we going to sell?)

By far the most common component of a campaign to be defined is the product to be sold. One reason is because upper management tends to view the marketing problems on a line-by-line basis.

Also, it is often the manager of a particular line of business who initiates the marketing process. For example, the manager of wireless services might be trying to respond to a particular erosion of the customer base.

Whatever the reason, the marketing team usually knows what product it is trying to sell as part of the initiation of the project. The only situation in which the product is not defined upfront is in the case of wallet share. In that case, the merchandise component of the campaign might not be selected until the very end of the development phase.

Media (How do we deliver the message?)

The next thing the marketers have to include in the marketing plan are the different media that they intend to use to deliver the marketing message.

Marketers have many tools at their disposal. They can use advertisements in many different forms (television, radio, and newsprint), direct marketing (telephone and mail), conferences, trade shows, public relations events, and a host of other techniques.

One of the most important delivery vehicles at the marketer's disposal is the telecommunications company itself. The marketer can include in the repertoire messages for the sales force to deliver regarding changes in pricing policies, input into the execution of customer service activities, and a long list of other "levers" that can be pulled. In most situations, the marketing team will have predetermined which particular medium or media will work best and will choose the media for a campaign during the first stage of the marketing process. At other times, the media will not be chosen until the modeling or development stages.

Sometimes, marketers choose the media for the campaign at the same time they choose the merchandise. This generally happens because someone has already determined that the particular medium will meet the organization's needs the best, or because the sponsor simply has a preference for that medium over others. Some sponsors want to work only with television. Some believe that direct marketing via outbound phone is best.

When the medium is chosen this early in the process, much of the subsequent campaign development work is easier.

Market (Whom will we sell to?)

Whom to send the marketing message to is a more complicated matter. During prioritization and goal setting, only the most generalized notions about who the target audience is for the campaign are developed. The bulk of the market segmentation work is done during the modeling phase. Prior to that, however, the marketing sponsor may have a good general idea about who should be targeted, or perhaps the results of a segmentation study are the reason that the campaign is being launched in the first place.

Messages (What will we say?)

Marketing messages define what the company will say to the prospect or customer. Marketing messages can be delivered directly through an explicit statement, such as "Use our phones, because we like you," or an implicit message, such as the one sent when monthly billing rates are lowered as a "reward" for being a good customer.

Sometimes a message will be developed during the prioritization and goal-setting stage, but more often than not, it is created during campaign development.

Margin (How much profit?)

Ironically, the last component of the campaign that we mention—and the one least often included in the decision-making process—is by far the most important one: For how much are we going to sell the service or product? How much profit will generating this new business mean to the company? The answer to these questions will also answer what is in every marketer's mind, namely: Is it worth it to run a particular campaign?

In reality, many telcos execute campaigns without any idea of whether they will be profitable. In those cases where margin is taken into account, the company will have a huge competitive advantage, because it will not spend money on campaigns that generate less than profitable customer relationships.

Prioritization and goal setting: interconnectivity

By the time the prioritization and goal-setting phase of the marketing process is completed, a team, a project, a set of objectives, and the general nature of the components of the campaign will have been mapped out. It is then the job of the team to carry this newly formed project through the modeling and campaign development phases.

From prioritization and goal setting to modeling

The transition from prioritization and goal setting into the modeling phase is usually a smooth and almost invisible one. At some point, the team will decide that enough conceptualizing, planning, consensus building, and decision making has been done, and that it is time to move ahead and begin analyzing some of the more critical components of the campaign and developing some specific models.

The information that moves forward through the process includes all of the decisions made by the team in its initial formulation.

From postcampaign analysis and feedback into prioritization and goal setting

Defining the order and the steps in the process is important because the marketing process is so dynamic and interactive that basically anything done anywhere can spur activity somewhere else. For example, how we describe the prioritization and goal-setting process here may look like some kind of "fresh start." In reality, marketers have access to a lot of feedback information that can help them make better decisions about project initiation. Some of the main sources of input to feed this step are discussed in the following subsections.

Previous campaign results

One of the most useful pieces of intelligence is information about how successful a previous campaign was. The more adept the organization is at measuring campaign results, the better they are at shortening the decision-making cycle and making effective decisions. For example, one successful campaign provides the marketer with much information for running the next campaign. If a newspaper ad in the sports section on Sunday yielded a far higher response than any other ads, then the marketer might build a campaign designed to target that particular segment.

Modeling-inspired new initiatives

Another valuable piece of information can come to the marketer through the efforts of modelers working on other projects. A modeler often discovers a valuable insight, trend, or perspective that inspires the marketer to launch an entire campaign.

For example, a modeler might discover that a certain kind of customer is most likely to be high-profit, low-maintenance, and low-risk. That could lead the marketer to launch a campaign to pursue specifically those customers.

15

The Modeling Process

I can predict the weather with 100% accuracy. All I need to know is the exact location and direction and speed of every single atom and sub-atomic particle in the entire universe at any given point in time. This, of course, is impossible, and so therefore, is the prediction of the future.

Albert Einstein (paraphrased)

The future is made up of the same stuff as the present.

Simone Wiel

Marketing projects and the modeling process

Once the project team has been formed and financed, the team sets out to investigate different ways to meet the sponsor's objectives. The development of the actual campaign, of course, requires the marketer to make many decisions. As the five components of the campaign are being defined, the marketer will want some assurance that the decisions being made are the correct ones:

- If I use television ads instead of newspaper ads, what will be the difference in response rates?

- Which medium delivers the best market penetration for each dollar spent?

- If I use a predatory campaign, instead of a loss-leader campaign, what will the impact be on the revenue generated?

- How will changing the price of the service offered change the number of people who sign up?

- Should I send these ads to young or older people?

Figuring out the answers to these questions is very difficult and the marketer will need some serious help in determining the best tradeoff decisions. This is exactly what the modeling process is all about.

It is the job of the modeler to help the marketer figure out which of the many hundreds of options that are available is the best choice for the particular campaign.

Marketing plan optimization: a multidimensional problem

Perhaps the marketing process we have described seems too simple. Management states a marketing objective, and the marketing team proposes one or two alternative approaches to meet those objectives. Simple enough? Not really! Let's see how complex the problem of putting together a proposal for this kind of activity can get.

A typical scenario

Management wants you to increase market share by 10%.

"No problem," says the naïve marketer. "We'll create an advertising campaign, put some ads on television, and wait for the new customers to come pouring in."

"How much money do you need for the advertising?" asks the management sponsor.

"Oh, four million dollars ought to be enough," responds the marketer.

The sponsor then asks, "How many new customers will this advertising attract? Will the number of new customers that we bring in offset the cost of doing the advertising? Will your ads disenfranchise some other groups of customers and actually cause us to lose market share? What if we cut the advertising budget in half? Will we get half as many new customers?"

"Oops," says the marketer. "I guess I have a few little details to figure out yet."

Applying mathematical and logical discipline to the marketing process

When a marketing team works on the development of an optimum marketing plan, the amount of work and the degree of sophistication expected are staggering.

A typical marketing project will often involve the development of dozens of alternative models in the investigation and support of various alternative solutions. Each of these models is developed to help the marketer build the ultimate plan that will be provided to management.

To make the marketing process work so that the marketer can make intelligent recommendations regarding what, when, where, and how to do it, an overall logical disciplined approach to the investigation of alternatives is necessary. We refer to this logically based discipline as the marketing modeling process.

Modeling: the optional process

One reason why the marketing process is so difficult to pin down and structure is because of the nature of the modeling phase of the process. You see, each of the other phases of the marketing process is required and must be executed in some kind of sequential order, but the modeling process is completely optional and can be invoked by almost anyone during any stage of the marketing cycle.

People perform modeling only when they think they need to—and they need to when there is some aspect of the nature of the campaign's definition that could be improved by the support of some quantitative, analytic type of input. If a marketer has a lot of experience, it is possible (and is often the case) for the marketer to skip modeling altogether and move from prioritization straight into campaign development.

Modeling as a legitimate process step

As just mentioned, marketers sometimes skip the modeling step altogether; however, certain key analytical modeling activities are commonly performed by marketers on a regular basis. These "regularly scheduled" modeling events are most often used to create information that will help the marketer prepare for the eventual campaign development process. This most common set of modeling functions is what we refer to as the *modeling phase* or *modeling step* within the overall marketing process.

What is modeling?

According to the dictionary, a model is "a description or analogy used to help visualize something which cannot be directly observed" (or, models are visualization tools) and "a system of postulates, data, and inferences presented as a mathematical description of an entity or state of affairs" (in other words, models are a mathematical description of reality used to help predict future events based on historical information).

In marketing, we use the modeling process to help us accomplish both.

Using models for visualization

One of the truly useful results of a modeling effort is that it presents a picture of the complex world of the customers and their behavior that helps everyone understand the marketplace and consumers' desires and demands. Modeling and segmentation do this job exceptionally well.

Using models for mathematically based marketing: quantification

Most people who use the term *modeling* in association with the marketing process are usually thinking about the word in terms of the second definition given above. The art and science of marketing during the twentieth century has developed a dazzling ability to predict customer behaviors and responses to a degree of accuracy never before imagined.

The modern marketing professional applies advanced statistical analysis techniques and complex behavioral and performance axioms to the problem of trying to understand when, where, how, and why people buy the products and services. With these tools, marketers can now develop models that allow them to provide the company with pinpoint accuracy in the development of their marketing campaigns. It is these kinds of models, the complex, multilayered, predictive models, that have made marketing such a powerful component of a corporate team.

Marketing model: a definition

A marketing model can be very large and complicated or relatively simple. It helps the marketing team provide management with a clear, accurate idea of what the consequences and impacts of a given marketing activity will be on the customer/prospect base. A formal definition of the marketing model follows:

> A marketing model is a logical collection of postulates and proofs assembled by a marketing team to describe and predict the customers' and prospects' responses to a specific component (or combination of components) of a given campaign proposal (message, media, market, margin, merchandise) with measurable and provable accuracy.

In general, therefore, a marketing model consists of a series of logically related assumptions, inferences, extrapolations, and theories, all tied together and proved with logically related models and reports.

Different types of models

Not only do models perform two major functions (visualization and quantification), but we have many different kinds of models available to us. In fact, models can be developed in support of each of the different major components of the campaign.

Market (segmentation) models

The segmentation-based model is the one that people most readily associate with marketing. Chapters 10, 11, and 12 focus on this kind of model development. Those models tell us about the customers themselves, their characteristics, and their behaviors.

Segmentation models are, however, only one small part of the overall modeling process.Understanding customers and their behaviors in a vacuum, without information and models to support the other four components of the campaign, is an almost meaningless exercise. It is the combination of segmentation models with the other types of models that makes the modeling process so powerful.

Merchandise (product) models

Product-based models include all of the operations and analyses that are developed to understand the viability, profitability, and utilization profiles for the different products that the telco has to offer. We talked briefly about this kind of analysis in Chapter 10.

Product models include the development of the different kinds of reports that allow the analyst to get a clear picture on how customers use products. Product models are particularly important in the development of wallet-share models, in which the marketer is trying to understand how multiple products are related to each other in the customer's mind. (See Chapter 8 for more information.)

Message (objective and approach) models

The message-based models are significantly different from other kinds of models. While the others are quantitatively-based and focused on the development of numbers and proofs for the justification of campaign decisions, message models help provide a vision and describe what the basic objectives and approach of the campaign will be.

We discussed many of the message-based models in Chapter 8. Some of the more popular include the following:

- *Acquisition message models:* startup, incremental, and predatory;

- *Retention message models:* buyer reinforcement, image enhancement, price adjustment, service level adjustment, "devil you know";

- *Wallet-share message models:* bundling, consumption enhancement, portfolio expansion.

These models provide almost no mathematical proof capability, but instead describe for the marketer a basic approach to be taken with the customer, which will then, in turn, proscribe a certain collection of the other types of models to make them achieve maximum efficiency.

Media (campaign performance) models

Another less than straightforward categorization for models can be found within the realm of the media models. Media models might be called *campaign performance* models for some obvious reasons.

Challenge of measuring performance

As you may recall from our earlier discussions about running campaigns via different media, one of the marketer's biggest challenges is figuring out how to measure the effectiveness of different campaigns. Because of the diverse nature of these media, different models must be developed to help the marketer predict how well the campaign will work.

Direct marketing performance models

The development of direct marketing performance models has refined itself over the years into the basic scoring and gains chart measuring approach. (See Chapters 6 and 13 for more details on these models.)

Mass marketing performance models

Mass marketing campaigns (television, radio, newspaper, and so on), on the other hand, have developed their rating and predicting systems on a totally different plane. The "price for points" model, which uses standardized demographics profiles and public penetration percentages (see Chapter 8 for details), is the fundamental model used in these cases.

Margin (profitability) models

Last but not least, is the requirement for the development of margin (profitability) information. As discussed in Chapter 8, these models include price elasticity, retention versus acquisition cost, price/utilization ratios, and other forms of analysis.

Hybrid models

While each of these forms of modeling has merit in and of itself, the real value comes from combining them to define the campaign parameters. When several modeling approaches are joined to create one discrete model, we refer to them as *hybrid* models. One of the most commonly used and widely defined models of this type is the customer value function (see Chapter 13). This function is really nothing more than a combination of the margin, merchandise, and market models for a given customer to produce one descriptive number.

Compound models versus discrete models

Until this point, we have referred to models and the modeling process as if they were some kind of singular event. In reality, however, the term *models* has two different meanings.

On the one hand, we have the concept of the model as it refers to the discrete development of a small, singly focused investigation of specific parts of an overall marketing plan. We refer to these individual modeling events as *discrete models*.

On the other hand, we have this concept of the overall process of collecting and executing a large number of small, discrete models, and combining them to create the large, multifaceted justifications that explain how different marketing activities should be pursued. The truly powerful models that come out of the modeling process are developed when marketers combine many different, logically related models into complex marketing megamodels. We refer to these as *compound models*.

Together, they form the basis for the formulation of a full-blown marketing plan. Discrete models are created often and for a variety of reasons. Compound modeling occurs when a marketing team assembles a logical collection of models into the justification for a specific marketing plan.

During the prioritization and goal-setting process, the marketing management staff, in conjunction with corporate executives and the operational managers, decide on the lines of business, the groups of customers, and the goals of the marketing teams. During the formal modeling process, however, these pieces are pulled together into a cohesive plan.

Components of a discrete quantitative model

To clarify the differences between discrete models and compound models, we begin by developing a better understanding of the discrete quantitative model.

As we can see from Figure 15.1, every discrete model has the following components:

- A set of data;

- An analytical operation run against the data;

- A hypothesis about the data;

- A conclusion that can be drawn based on the results of the analysis.

Let's see how this works in real terms.

Segmentation model example

A marketer has decided to learn more about the customers. The objective of the marketer is to determine the different major groupings of customers. To develop this discrete model, then, the marketer might do the following:

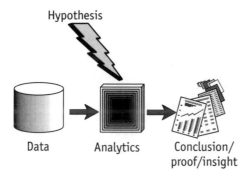

Hypothesis

Data   Analytics   Conclusion/
                                 proof/insight

Figure 15.1 Components of a discrete quantitative model.

1. *Data.* To do the analysis, the marketer needs data. In this case, the data will probably be information about existing customers' personal characteristics.

2. *Hypothesis.* An applicable hypothesis would be that "there are different major groupings of customers within this population that can be used to help develop more effective campaigns."

3. *Analytical operation.* Based on the data and the hypothesis, the modeler chooses an analytical tool (say, a statistical analysis package like SPSS) and decides which kind of analytical operation to run (perhaps a cluster analysis). After that, the modeler runs the analytical operation and creates output reports.

4. *Conclusions.* Those reports allow the analyst and marketer to jointly interpret the results and decide (a) whether the hypothesis is true (Was a useful grouping found?) and (b) if yes, what are they and how can they be used?

These basic components and basic steps define what a discrete quantitative model is and how it is run. The same basic approach applies to running a market (segmentation), margin (price/profit), merchandise (product), or media performance model.

Modeling "chains"

A single discrete model may provide the marketer with some small piece of useful information. However, what is more interesting is the cumulative effect that running dozens or even hundreds of discrete models can have on the marketer's understanding of the different alternatives available in the development of a campaign. There are literally hundreds of decisions to be made and the right series of discrete models can help narrow down the decisions tremendously.

Although it is possible for a marketer and an analyst to sit down and figure out what the dozens or hundreds of different discrete models might be that will shore up the campaign decision-making process, a more dynamic and interactive technique has been developed over the years. This technique is referred to as *model chaining*.

Model chains: a definition

A modeling chain is a logically related series of discrete quantitative models, each of which is used to provide input into the link that follows it and which is fed by the link before.

The chaining of discrete models allows the knowledge learned from one model to affect and feed the decision-making process of the next one.

As we can see in Figure 15.2, the first model was run, and some conclusions were developed. By examining the outputs of the model, the modeler will then be able to:

- Develop the next hypothesis.

- Decide what new data might be required to test the new hypothesis.

- Determine what form of analytic to use to test the new hypothesis.

As the marketer gets better at building these modeling chains, the whole process becomes faster and more accurate.

Keys to successful modeling

The modeling process and the development of modeling chains is the primary reason why analytic marketing has become so popular and so successful. It is an incredibly powerful, flexible, and insightful tool. To make the process work, however, you need three very important things:

1. Marketing and analytical personnel who are capable of building, running, and managing this process;

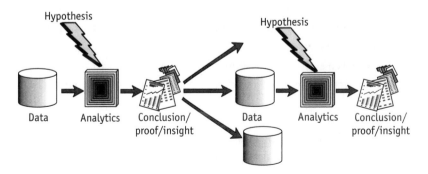

Figure 15.2 Two links in a modeling chain.

2. Analytical tools that make the modeling easy to do;

3. Access to lots and lots of difficult-to-manage, hard-to-find data.

Role of data in modeling chains

Whereas the right people, the right tools, and the right ideas are certainly important to the support of the modeling process, none of them will be of any value if the modeler does not have access to the right population of supporting data.

Although it is certainly true that a well-run marketing database (data warehousing/data mining) environment can make the job of the modeler easier, it would be a big mistake to think that it is possible to pre-define all of the data requirements that this process will need.

The process of data model chaining is incredibly dynamic, and the source of its greatest power is that the modeler can pull in previously unconsidered populations of data for inclusion in the analytical work.

Because of this, the construction of a data model chaining development environment is a unique and painstaking portion of the overall marketing database environment.

Modeling during the modeling step

The majority of modeling is done during the modeling step itself. In other words, after a marketing team is formed and objectives defined, the team will proceed to do some modeling work.

In general, modeling during the modeling phase concentrates on developing segmentation models. Direct marketers use the segmentation information to choose base lists for campaign development and mass marketers use that information to define the media, events, and the messages they want to convey.

Coupled with this basic segmentation work is often some related modeling efforts in the merchandise (product) and margin (profitability) areas.

Special role of message models

Whereas the actual series of model chains and the interplay of different market, merchandise, and margin discrete models within those models will vary greatly from campaign to campaign, the process is not quite as chaotic as it might first appear. Most marketing organizations find that each major message model, such as predatory, bundling, and startup, has associated with it a predefined common set of model chains, traditional combinations of market, merchandise, and margin models, which are organized so that they logically support the development of those kinds of campaigns. We refer to these message model-based modeling chains as marketing model *templates*.

Modeling during the campaign development step

The other major use of models occurs during the campaign development stage when the media performance measurement models are employed. During this time, scoring and price-point analysis types of operations are performed to ensure that the campaigns will deliver the desired results. Another key element of this kind of model is the testing component. (See Chapter 16 for details.)

Marketing model templates

The process of building these large, complex models can be very confusing if they are not structured with some kind of discipline. The structure for the development of these models, in fact, springs from two basic premises.

First, the marketing models for the majority of marketing activities can be boiled down to three basic approaches, namely, acquisition, wallet share, and retention, and the different types of marketing messages that they deliver.

Second, while the details of how the models are put together will vary tremendously from one project to the next, they will all follow the same basic logical template or series of assumptions and proofs to get the job done.

Let's take a look at some examples.

Logical templates: key concepts

The major building blocks of the logical template are the following concepts:

- *P:* describes the population of people we are working with;

- P_{CUST}: represents the population of all customers;

- P_{PROS}: represents the population of all prospects (people who are not yet customers);

- *B:* describes the behavior function of importance for a selected population;

- B_{CUST}, B_{PROS}: represent the behavior functions for customers and prospects, respectively;

- *C:* represents the characteristics (nonbehavior attributes) of the population;

- C_{CUST}, C_{PROS} : represent the characteristics for customers and prospects, respectively;

Finally we use the mathematical convention of "tick marks" (primes) to represent the related subsets of these populations. Therefore, P_{CUST}' (pronounced PCUST-prime) represents a specific subset of customers in the population, and P_{CUST}'' (pronounced (PCUST-double prime) represents a different subset of customers.

Acquisition (preferred behavior) model template

To show how a specific message model can be used to dictate the construction of a modeling chain, we will look at one particular case, an acquisition campaign to attract more customers with the most desirable behavior. (In other words, we want to attract only the best customers.)

Basic assumptions

Our assumptions around the development of this acquisition model are as follows:

1. The marketing team has determined that the objectives set by the sponsor will best be met, either fully or in part, by the addition of more customers to the company's ranks.

2. The main focus of the model will, therefore, be to concentrate on the customers' subscribe behavior, rather than the utilization behavior.

3. If the team determines that it needs to include both subscribe behavior and spending behavior in the model, then it has to combine the acquisition and wallet-share models. (At this point we assume that our goals will be met by a simple acquisition.)

4. We also assume that the objectives of the team will be best met by the attraction of certain kinds of customers. (Our case is not a market situation in which the acquisition of all customers is the objective, regardless of how good or bad they might be. These situations actually do occur, especially in new markets where there is little or no competition.)

Basic components

Given that these criteria have been met, we are now ready to consider exactly what components we need to develop our acquisition model. Remember, the purpose of the model is to identify the components of a marketing plan that will meet the objectives of management the most economically. That plan will consist of a segment, a media, a message, a price (sometimes), a time frame, and a product or service to be sold.

To develop this model, we need the following kind of support data:

1. Information about the profitability of the product to be sold and the customers' elasticity regarding different prices at which it might be offered;

2. Information about the characteristics of current customers with the behavior or profile that interests us;

3. Information about the capacity of the organization to meet the new demand;

4. Information about the different groups of people to which the product might be marketed (information gained from census reports, purchased lists, purchased market intelligence, organizationally run surveys or questionnaires, and a variety of other sources).

Armed with this background information, the marketer is able to build the acquisition model. If key pieces of this information are missing, then the model cannot be run or will have some serious flaws in its logic (with associated risks).

Contributory models

Provided with this kind of background information, marketers will make use of a variety of modeling techniques including (1) product profitability reporting, (2) segmentation, (3) list scoring, (4) credit scoring, and possibly some others, depending on the model.

If the organization has gone through the trouble of developing a customer value function, it could be used for the model, eliminating the need to do all of that work.

The acquisition model: logical template

The easiest of the logical model templates to follow is probably the one for acquisition activities and we will therefore take a closer look at it. Remember that a model's logical template shows us, at a high level, how we develop a proposal for spending money on the creation of a marketing campaign. Campaigns are defined as sending a particular message to a particular group of people (segments), through a specific vehicle (media), within a given time frame, for a specific product line.

The objective of this template is to show us how we can predict for management what the impact of this campaign will be on revenue, market share, and profit before any money is spent.

Once this model has been developed, management may ask us to "tweak" the model and change parameters, and then report on the effects of those changes.

Step 1: Review product profitability reports

The first step in the preparation of this kind of model is for the marketer to review the profitability, price points, and customer price elasticity associated with the product to be promoted. Reports like these should be available for the profitability part, and marketers can do some quick price elasticity analysis of their own purchasing file in combination with a review of the competition's price position in order to get a general idea about price elasticity. With this information, the decision can be made whether or not the product should be promoted and under what conditions.

Step 2: Identify the desired behavior

The second step in the process is to identify which behaviors we want to see exhibited by the customers we attract.

We start with the entire population of customers for the given target area (e.g., product line, segment) P_{CUST}, with a wide variety of different behavior functions, B_{CUST}. Then we assume that within that population of customers, there are some people with a behavior that is preferred over others. For example, if the objective of our campaign is profit, then the people who spend more money on the profitable product lines exhibit the desired behavior. If, on the other hand, our goal is to maximize revenue, then we look for customers who utilize more of our services.

Equally important at this point is to identify those individuals with behaviors that we do not want included in our new customer list. We definitely want to avoid customers who are expensive to keep, unprofitable, difficult to maintain, and negligent about paying their bills.

We assemble this collection of desirable and undesirable behaviors and use it to identify the subset of current customers we would like to see many more of. These customers with the preferred behavior function are referred to as the B_{CUST}' (BCUST-prime).

Step 3: Identify the characteristics of preferred customers

Once we know the B_{CUST}', we return to our customer population P_{CUST} for further analysis. We need to identify those customers within our overall population who exhibit the behavior function we are interested in, B_{CUST}'. This provides us with a new subset of our entire population, namely P_{CUST}' or the subset of preferred customers.

After that, we figure out what characteristics this particular subset of customers shares. For example, are all of the preferred customers of the same age? Do they share political beliefs? Do they live in the same city? In other words, we need to perform a detailed segmentation analysis of this subset and look for truly indicative collections of characteristics.

Step 4: Identify a population of prospects having the same characteristics

Next, the marketer turns to the general population of noncustomers P_{PROS}. The objective is to find those individuals who have the same characteristics as the people in the preferred customer population C_{CUST}'. When $C_{CUST}' = C_{PROS}$, we have identified a new population of prospects B_{PROS}' whom we believe are similar to our preferred customers.

Step 5: Inference of prospect behavior

Our model says that we should focus our attention on attracting only prospects who we believe will exhibit the behaviors that we want to see in our customers. Since the identified subset of prospects has the same characteristics as the preferred subset of customers, we can infer that the behavior of the prospects will be similar to that of the preferred customers, if we can convince them to become customers too.

Furthermore, we should be able to carry the inference even further and assume that if we attract similar kinds of customers, they will probably spend the same amounts of money as our existing customers.

Step 6: Identify the best way to reach the targeted segment

Once a target group of prospects is identified, the marketer can then figure out which media will do the best job of reaching them. If the segment is very small, or very specialized, then the direct approaches are preferred. If the segment is large or homogeneous, then mass marketing approaches are likely to be more successful.

Of course, an infinite number of ways exist in which mass marketing can be targeted (using focused messages and limited mass media, such as magazines) and there are many situations where direct marketing might be used in a mass marketing mode.

In all cases, the media must be selected before the modeling can continue.

Step 7: Determine how to make the offer

After choosing the segment and the media, the marketer can concentrate on the development of a specific message. The message will be unique to the campaign, the segment, and the media being used.

Step 8: Test assumptions

In the best cases, the marketer will test the message on a sampling of the segment (through small sample mailings, use of focus groups, opinion polls, and so forth). When no testing is done, the marketer has to make up test assumptions based on experience or just plain intuition.

Step 9: Calculate campaign values

Now, the marketer is ready to appraise the overall economic viability of the campaign by developing scoring models similar to the ones used for list scoring.

For direct marketing, actual list scoring models are created. For mass marketing, segment-based scoring models that assign probabilities and response rates to the behaviors of different segments are used.

In all cases, the marketer is left with a comprehensive, mathematically based case for moving forward with the campaign.

Step 10: Mass marketing

If mass marketing is chosen as the medium, the marketer must map to social class and demographics (mass media target domain metrics).

Step 11: Measuring results

Of course, we all know that experience is the best teacher, and this is certainly true in the case of modeling. The real key to making the modeling process work is for marketer to stay with the campaign after it is run and determine exactly how effective it was. In this way, we learn from the experience.

If the results are exactly as predicted, then the marketer can be congratulated. If the results were higher or lower than expected, then the marketer needs to reevaluate the model for faulty assumptions.

In an environment where marketers can keep track of what they are doing, the cost of developing models tends to decrease and the quality of those models tends to increase exponentially.

Variations in the process

The model development process we have outlined is an idealized description of how modeling is done. In reality, of course, many adjustments will have to be made.

16

Campaign Development and Execution

The reward of a thing well done, is to have done it.

Ralph Waldo Emerson

In action, be primitive, in foresight, a strategist.

Rene Char

When action grows unprofitable, gather information; when information grows unprofitable, sleep.

Ursula K. Le Guin

Act quickly, think slowly.

Greek proverb

At this point in the life cycle of the marketing campaign, we come to the end of the planning, thinking, and talking stages and move into action. The emphasis of the prioritization and goal-setting and modeling phases is to help the marketer make the right decisions about the campaigns to be executed. In reality, however, we are already putting those insights into action by the time we get into the development and execution stages.

Marketing projects and campaign development

When the marketing project leaves the modeling stage, the team has completed a marketing plan, which identifies at least parts of the message, medium, market segment, merchandise (product), and margin (price). In addition, a model has been developed that provides management with a rationally and statistically based approximation of what the effect of this campaign will be and what it will cost. Once the model and plan are approved, the team is ready to move the project into the campaign development phase.

During campaign development, the team expands its membership to include different groups of people, depending on the media. For mass marketing, advertising agencies are usually employed. For direct marketing, a direct marketing specialist is engaged, either from within the firm or from outside.

Objectives of the campaign development stage

At this stage in the marketing cycle, the marketer is concerned with finalizing the campaign component decisions and getting the actual ads ready for execution. The marketer may add the step of running tests prior to making a final commitment. These three tasks are discussed next.

Finalize campaign component decisions

Up to this point, the marketer has been able to postpone making any of the really tough decisions about how the actual campaign will be put together. Indeed, the prioritization and modeling processes are geared toward allowing the marketer to be creative and explore the consequences of different campaign component options. When we move into campaign development, however, the time for experimentation has passed. Now, the marketer must make hard decisions and finalize all the different characteristics of the campaign.

Here is where the discipline and experience of the marketer comes into play. If the marketer has been in control of the marketing process, he or she will enter this phase with confidence, knowing what needs to be done and how to get it done in the time allotted. On the other hand, if the marketer has been naïve or undisciplined, then this phase will be characterized by a lot of panic and rash decisions that may turn out okay if the marketer's instincts are good, but could also turn out to be disastrous.

Run tests (optional)

One of the main remedies available to the marketer when the campaign is less than optimally assembled is to take the campaign for a "test drive" before committing the organization's resources. Running tests can help the marketer solidify the decisions about the five aspects of the campaign, and provide a good initial "read" on how well the campaign will do.

Testing has been the salvation of many marketers, allowing them to pull out of campaign commitments before major blunders were made.

Prepare materials for distribution

After the details about the five components of the campaign are finalized, the marketer faces the painstaking task of actually assembling and coordinating the many different details that make up the actual guts of the campaign itself.

Here too, the efficiency of the marketing process and the competence of the overall marketing organization will be clear. Literally hundreds and possibly even thousands of details must be addressed when preparing any kind of campaign, and the marketer must double and triple check all of them—or risk potential disaster.

Major steps in development

Now let's look at the three basic steps that the marketer will go through during campaign development in a little more detail.

Finalizing the five components

Of course, marketers cannot proceed with a campaign if they are not sure about the market, message, media, margin, or merchandise to be delivered. By this stage, the decision about each component will fall into one of three categories:

1. Some decisions have already been made simply because they were obvious or had been decided ahead of time. For example, the sponsor may have specified that a campaign be for long-distance service (merchandise), that a television ad be used (media), and that no special price offers be made or mentioned (margin).

2. Some decisions have been investigated through the development of supporting models and have been chosen based on that evidence. (Market segment decisions are usually made this way.)

3. Some decisions have not yet been made. Development of a specific message, for example, might be postponed until the end.

Obviously, in different situations, different components will fall under these three categories. Sometimes, for instance, the media decision is postponed until the end. Other times, with bundling and price adjustment campaigns for example, margin (price) decisions are postponed. In almost all situations, however, some aspect of the campaign has yet to be resolved.

Finalizing decisions and the role of testing

When marketers find that despite all of their best modeling efforts there are still just too many questions about the right combination of elements, the ultimate place they can turn to get those issues resolved is to testing. When decisions still need to be made that have large risks associated with them, the marketer will often choose to conduct formal tests.

Testing

The role and the value that testing has for the marketer cannot be overemphasized. Although disciplined, quantitatively based modeling efforts can certainly contribute to the marketer's understanding of people and their potential reactions to different marketing activities, there is just no substitute for a real trial run of the campaign to give the marketer some hard experience regarding the campaign's ultimate effects.

Reasons for testing

Running tests is useful for a number of reasons. Some of the more prominent reasons include these:

- *Gathering input for modeling efforts.* Small tests can be run to help marketers gain the data needed to better feed the modeling process itself. For example, if marketers are not sure what the demographic makeup of a group of customers is, they might test market to a small subset of that group and use the feedback to better fill out their modeling assumptions.

- *Validating assumptions of conclusions developed as a result of modeling efforts.* One of the main reasons tests are run is to provide the marketer with evidence to back up the conclusions developed during the modeling process.

- *Assessing the viability and effectiveness of messages.* By far the biggest reason that marketers run tests is to find out if customers will react to messages as anticipated.

- *Gathering hard data regarding the campaign's probable return on investment (media models).* Finally, and as a combination of all of the others, marketers run tests to help them develop the media models (scoring, gains charts, market penetration, and so on) that allow them to decide which medium to use and the expected return that media investment will have.

Types of tests

With such a wide variety of choices for different media, messages, and reasons for testing, it makes sense that there would also be a wide variety of tests to run. A few of the different types include the following:

- *Geographic isolation tests (mass market).* One big challenge of the mass marketer is to figure out how to test a small sample of the population without exposing the entire population to the message. The easiest way is to pick a limited geographic section and execute the entire campaign in that area only. In this way, the impact of the campaign will be limited to one region until the effects of the test can be evaluated.

- *Random sample tests (direct market).* The direct marketing approach can be used to pinpoint individual customers. This pinpoint accuracy means that the marketer can select a random sample of an entire population and test the campaign message in a limited way. The results from the small test will give the marketer a strong indication of how well the overall campaign will do.

- *Segment tests (direct and mass market).* Another approach available to both mass and direct marketers is to run the test with only one segment or subsegment of the entire population. The effectiveness of the campaign across a small subset can then be used to gauge the campaign's overall effectiveness.

- *Focus group tests (direct and mass market).* An approach used in the past almost exclusively by the television media, but now becoming popular with all media testers, is the focus group. Focus groups, test panels, and other testing approaches involve gathering volunteers. This group is then asked to review different advertising materials and to explain how the materials make them feel and what kinds of actions they might encourage.

The role that tests play in the development of a campaign will vary greatly depending on the nature of the campaign (Is it a new, untested concept or approach?) and the size of the investment (you are much more likely to test before running a $5 million campaign than a $50,000 campaign).

The media production cycle

Eventually, the marketing team settles on all of the different components that will make up the campaign, and the marketer is ready to order the actual preparation of the materials. At this point, the marketer takes a back seat and the media production specialists come into play.

Developing advertising material for each of the different media discussed is a diverse, complex, and expensive process, and best left to the professional to manage. The following sections provide a general idea about the different steps involved in the preparation of each of the different kinds of campaigns. Large telecommunications firms generally have entire departments dedicated to managing these kinds of operations. In the case of mass media, these departments concentrate on managing negotiations with advertising agencies and supervising their activities.

In the case of direct marketing, many large telcos have their own outbound phone and in-house print shop capabilities. As the size of the telco decreases, however, more and more of the diverse marketing tasks become the responsibility of fewer and fewer people.

Television preparation

Preparation of television ads involves the following production steps:

1. *Storyboards.* Sketchboard mock-ups of the individual scenes of the commercial are presented to the marketer for initial conceptual approval.

2. *Scripts.* Detailed scripts that lay out the precise dialog, sound, and effects for the commercial are prepared.

3. *Facilities.* Sound, video, special effects, and related facilities must be located, contracted, and scheduled.

4. *Production crew.* A production crew consisting of producer; director; actors; camera, sound, and prop crews; and so on must be selected and employed.

5. *Rehearsals.* The production cast is often assembled for rehearsal before the actual shoot.

6. *Editing.* After shooting the commercial, sound and film editors are employed to assemble and produce the finished product.

7. *Scheduling and price negotiations.* Last, but not least, the actual time and place for the ad to run must be negotiated and scheduled.

Print media preparation

Preparation of print media, though not as involved as creating a video ad, is complex in its own right.

1. *Sketches and concepts.* First, initial sketches and concepts for the ads are developed.

2. *Drafts and approvals.* Once approved, any number of additional drafts will be submitted before artist and marketer are satisfied with the results.

3. *Typesetting and proofreading.* The approved material is then submitted to typesetters for casting and the final proofs are submitted to the marketer for approval.

4. *Scheduling, placements, and final proofs.* Finally, the scheduling of the ad, its placement within the publication, and the final proofs are developed.

Direct mail preparation

The direct mail campaign has a unique preparation cycle of its own and includes these steps:

1. *Sketches and concepts.* The development of sketches and concepts is similar to that for print media.

2. *Drafts, layouts, and approvals.* Drafting and approvals occur just as for print media.

3. *Typesetting and proofreading.* The same typesetting operations and proofreading as are used for print media occur, only this time they are for freestanding documents, not part of a bigger publication.

4. *Print scheduling and final proofs.* The final print schedule is determined.

5. *List selection, scoring, and production.* At the same time as this process is going on, the direct marketer will be doing all of the list preparation work: choosing base lists, scoring them, possibly testing them, and finally turning the ultimate (short) list of chosen recipients over to the direct mailer.

6. *Stuffing, collating, and mailing schedules.* Not only must the list be created and the material printed, but envelopes must be addressed, stuffed, and the postage scheduled for distribution.

Direct phone preparation

Direct phone campaigns also have a unique production schedule:

1. *Script preparation.* The first and most critical part of a phone campaign is the development of the script that will be used to solicit the customers. Scripts often undergo several rewrites while the optimum approach is developed.

2. *List selection, scoring and production.* This process is the same for direct phone campaigns as it is for mail campaigns. Lists are selected, scored, tested, and submitted.

3. *Scheduling and timing approvals.* Finally, the finished scripts and lists are submitted to the "phone room" for execution.

Marketing projects and campaign execution

After undergoing development and testing, the marketing plan, which is now associated with a specific campaign, is ready to execute. The execution, in most situations, will involve actually running the advertisements and mailing or calling the customers.

Objectives of the campaign execution stage

The campaign will have been finalized, tested, produced, and scheduled. Now we can actually run the campaign. For the marketer, this is an experience very similar to that of a performer on opening night. Everything thus far has been preparation, rehearsal, and more preparation. Now, however, the time has come to actually get on stage and see how the audience is going to respond.

Seasoned marketers know that this is not the time to sit back and relax. There are still several important things to be done, including these:

- Double check all details before execution.

- Keep track of the execution of the campaign.

- Monitor the short- and long-term effects of the campaign.

- Intervene when execution is going poorly.

When the campaign does not go well

We would like to believe that we live in a perfect world, but something almost always goes wrong and when that happens, the marketer should be in a position to step in.

Mechanical/execution problems

Some of the most frustrating and embarrassing problems that marketers run into when executing campaigns occur when things go wrong mechanically. Here are just some of the more painful experiences that marketers have run into in the past:

- Print ads were run in multiple cities and the phone numbers for customers to call were switched.

- Ads were run with the wrong soundtracks.

- Radio advertisers read the wrong commercials at the wrong time.

And many, many more.

Customer reaction/feedback problems

What if everything runs exactly as planned, but the customers react in unpredictable ways? Sometimes, especially when no testing is done, an ad will end up having an unexpected and sometimes even a negative effect. The history of cross-cultural advertising is full of examples of situations like this. The use of certain colors, certain attitudes, and certain words or names can create disastrous effects if not handled carefully.

When campaigns have negative impacts, it is critical that the marketer become aware of that as soon as possible and make immediate plans to cancel the rest of the campaign's execution if necessary.

Steps in campaign execution

Campaign execution, then, is much more about monitoring and possibly stopping the execution of a campaign than anything else. The steps involved with running a campaign are discussed next.

Prepare campaign monitoring mechanisms

Keeping track of ads and marketing activities is actually much more complicated than you might imagine. For example, if your company is sending out millions of letters, making hundreds of thousands of phone calls, placing ads in dozens of newspapers, and advertising on four different television stations, how can the marketer possibly hope to keep track of them all?

Assigning personnel to monitoring duty

One approach is to assign different members of the team or local employees who are part of the management, sales, or clerical staff to look for and give feedback about ads as part of their job responsibilities.

Timed rollout of campaigns

In the case of direct marketing, the alert marketer will have automatic feedback reporting mechanisms built in. A smart direct marketer does not execute the full campaign all at once, but instead schedules a gradual rollout so that early returns can be monitored and marketing disasters averted.

For example, to execute a one million name mailing, the marketer should schedule the mailing of 100,000 pieces every two weeks. This way, the marketer can see how well the first 100,000 did before committing to mailing the second 100,000.

In the same way, mass marketers can spread their ads out across an extended period of time, allowing them to measure results early and often.

High-volume, high-quality monitoring

For those organizations running an especially large volume of ads, specialized advertising tracking and monitoring organizations exist:

- *Clipping services.* Clipping services scan newspapers in local markets and actually cut out copies of your ads and send them to you for confirmation that they did indeed run when and where they were supposed to.

- *Media ratings companies.* Rating organizations such as A. C. Nielsen, through a variety of mechanisms, keep track of advertisements and measure the impact they had on the consumers that saw them.

Monitor immediate results

The first mission of the marketer, then, is to monitor the immediate effects that a campaign is having and to be ready to intervene if things go wrong. These are the mechanisms and situations we have already talked about.

Monitor ongoing results

The other job of the marketer, however, is to "close the loop" on the overall marketing process and to make sure that the long-term effects of the marketing efforts are being measured and fed back into the marketing process. When the process starts all over again, everything learned from the last campaign is included in the making of the next marketing decision.

Marketing project feedback and analysis

We have now come full circle: from prioritization and goal setting, modeling, campaign development, campaign execution, and back again into the prioritization and goal-setting cycle for the next round of campaigns. Throughout our examination of this process, we have tried to make certain key concepts clear and reinforce them along the way.

The dynamic nature of the telco marketing process

By this time you should have realized that the telco marketing process is extremely complex and dynamic in nature. At each step along the way in the development of a campaign, the marketer is faced with an unbelievable number of options and parameters from which decisions need to be made. Each decision, each subtle change of information can cause the marketer to abandon a current course of action and start an entirely new line of reasoning.

Interplay of information and dependencies between stages

At the same time that all of the dynamism is going on, the marketer is trying desperately to gather as much information as possible. Each stage of the marketing cycle develops its own kind of information and intelligence, and everyone working on other phases of the cycle can use the information that is constantly being generated.

Core competency holds the process together

Underlying all of the seeming pandemonium of a marketing campaign are a discipline and a core competency that make the whole process work. It is the marketers' ability to work in this kind of dynamic, unstructured, and creative environment that makes them a valuable resource to the corporation.

Core computer systems/marketing database capability

At the same time, marketers are unable or greatly curtailed in their ability to do a good job if access to information is not possible. Information is the lifeblood of marketing and it is the marketing database system and the core competencies that we have been discussing that make it all work.

Part 5

The Marketing Database

17

Marketing Database Functionality

Form can never supersede function.
Unknown

What exactly does this thing do?
Everyone

We have spent much time talking about marketing processes and functions and how the marketer can use them to help drive up company profits and increase market share. What we have yet to discuss is a very important part that has only been assumed thus far. That is the part played in this environment by the marketing database itself. Understanding the different types of marketing analytical techniques and how they fit into the way the telecommunications company runs is important, but those methods cannot be used without the correct system in place.

271

Most books about marketing strategies and marketing databases in telecommunications start with a detailed discussion of hardware, software, databases, and data mining tools. Those are certainly important, but to define how a good marketing database environment should be constructed, we need to be clear on how the marketer will use it. Because the physical structure of the marketing database is greatly dependent on the organizational and functional structure that the company imposes on it, it is critical that we first understand this functionality.

Basic functions

An organization expects the marketing database to address three core functional areas:

1. *Query and reporting:* allowing marketers and members of the operational units immediate access to information about customers, segments, products, and other information of interest;

2. *Analytics:* enabling marketers and members of the marketing project team to do sophisticated statistical analyses, data mining, and other forms of analysis;

3. *Process management:* providing the entire organization with software that structures and tracks the marketing process, including many of the core processes that drive marketing (campaign, list, customer information, segment, planning, and model management).

Each of these three areas is related to the others and shares much of the same underlying data.

Relating the database to the process

You may recall that the marketing process has four major functions: prioritization and goal setting, modeling, campaign planning, and campaign execution. During the prioritization and goal-setting step, the organization chooses sponsors, assembles teams, and sets goals. Let's see where each of these three functional areas comes into play.

Query and reporting subsystem

The query and reporting subsystem is by far the most widely used. It includes the technologies, formats, information databases, and functional contributions to each of the four marketing process areas.

Query and reporting technologies

To deliver the required information to users, the developers of a marketing database system need to include one or several different query and reporting tools. We review the more important types of tools and provide examples of some of the more popular choices.

Basic spreadsheets with database access

By far the number one tool preferred by marketers is the old reliable spreadsheet. Today's spreadsheets, predominantly Microsoft Excel and Lotus 1-2-3, provide marketers with many of the capabilities they need to work with the marketing database. Spreadsheets are not difficult to use, most corporate users are familiar with them, and their end products are easily shared among people. Most modern network configurations allow users to hook up their spreadsheets directly to a database with object database connectivity (ODBC) or some other intermediary protocol.

On the down side, spreadsheets have only limited query capabilities and place constraints on the size of files.

Basic report writer and query tools

The second most common form of query support is delivered by what are called *basic query* or *report writer* technologies. This set of tools includes products such as IBM's QMF and Oracle's ISQL. With these products, users can access and analyze much larger data files than with spreadsheets, but formatting and data manipulation capabilities are usually severely limited.

Advanced query/reporting tools

A whole new category of query and reporting tools has recently come on the scene. These products, such as Cognos, Business Objects, and many others, offer all the data volume capabilities of the query tools, while also delivering much of the power and flexibility of spreadsheets. They can often be used to write semi-OLTP type applications with a minimum of hassle. However, these products tend to be expensive and require a considerable amount of training.

OLAP technology

The OLAP tools are the latest, greatest, and most popular of the query and reporting toolset. They provide for dynamic "slicing and dicing" and navigating through large data stores in a relatively straightforward manner, allowing users a view of the relationships they would not otherwise be able to see. In fact, I believe that, given a choice, users would obtain all of their information through OLAP reports.

However, OLAP tools have three significant disadvantages:

1. They are difficult and expensive to set up, requiring many hours of specialists' time.

2. Because of the complicated setup, they are also difficult to change. Users must thoroughly think through their designs before OLAP cubes are built, because rebuilding is expensive and time consuming.

3. OLAP systems are limited by the size of data files they can manage effectively. Current technology precludes the use of OLAP for very large databases (such as call detail record systems).

Hybrids

Most products in these categories combine several different types. It is not unusual to see spreadsheet, query, and OLAP capabilities built into one product, making it extremely appealing to some organizations.

Formats

Output from these products can be as simple as a list of names or as complex as cross-tab reports, control-break reports, or OLAP systems. (See Chapter 9 for a review of these output types.) In general, however, these reports meet the customer's specifications, look good, and communicate information well.

Information databases

Users not only have a wide variety of tools at their disposal in this category, but usually also have an extensive amount of data from which to choose. They can select from customer views, product views, and process views.

Customer views

The most important set of data stores and the most commonly accessed by the query and reporting tools is the customer-based data. After all, marketing is about customers and prospective customers. Some of the primary data stores include the following:

- *Customer master file:* customer name, address, phone numbers, products subscribed to, and so on.

- *Customer financial history:* tracks the revenue a customer generates, the fees paid, and so on.

- *Customer utilization history:* information about how much of what products or services the customer uses.

- *Customer contact history:* list of the different ways customers have been contacted. (This consolidates sales, customer service, billing, credit, and marketing contact information; see Customer Relationship Management Systems in Chapter 4 for more information about this kind of system.)

- *Customer process view:* the various marketing processes the customers have been involved in, the segment(s) they were assigned to, their scores and rankings, and so on.

Product views

The query and reporting system also provides a product view that includes each line of business the telco covers (such as long distance, wireless, wireline, ISDN, T1, Internet, cable) and insights along the following dimensions:

- *Product utilization:* information about how much of each product is consumed by the minute, day, week, month, quarter, and year, in terms of geography, market, and other categorizations;

■ *Product revenue/profitability:* information about how much revenue these different products bring in and how profitable they are according to the same reporting criteria.

Process views

Query and reporting systems are also used to help the organization keep track of the different processes it manages, including tracking of campaigns, models, plans, and lists.

Query and reporting support skills

The skills required to make query and reporting viable in an organization vary greatly depending on the tools in use and the complexity of the database structures.

1. *Data analysis and building databases.* Having people on staff who know how to build the table structures is the secret to effective marketing databases. It is an extremely valuable and sorely under-recognized skill. For many organizations, 80% of the cost of building a marketing database is in finding the data required, formatting it, and storing it into tables for use.

2. *DBA skills.* The organization also needs a database administrator (DBA) who can take the stored information and turn it into a table structure that is easy for the users to access.

3. *User skill 1—knowing the data stores.* Most users have concerns about how difficult it is to use a query or reporting tool, but the real job they have to master is learning what is in each data store and interpreting the data.

4. *User skill 2—knowing the query/reporting tools.* Ultimately, of course, the user must learn how to use the tool itself.

5. *Expert support versus expert users.* In some organizations, the users themselves learn how to master the knowledge about data stores and query tools. In others, *help desk specialists* or *decision support specialists* are recruited to ease the burden on the users.

Marketing steps and query and reporting contribution

The query and reporting system is the single most widely used aspect of a marketing database system. It supports the execution of each step in the marketing process (prioritization and goal setting, modeling, campaign planning, and campaign execution). Equally important is the fact that this system shares the information that marketing has about customers, products, and processes with any user in the corporation that needs it, namely, corporate executives and members of operational units such as finance, accounting, operations, and sales (see Figure 17.1).

Analytics

Although the query and reporting subsystem is the most important and widely used part of a marketing database environment, most people associate the term *marketing database* with the analytics portion of it. Analytics, the application of mathematical disciplines to the solution of marketing problems, is clearly the place where modern marketing gets

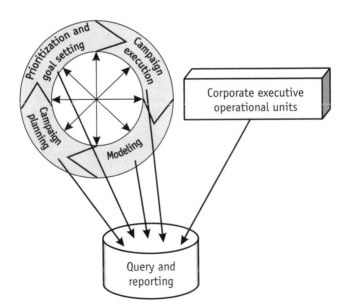

Figure 17.1   The marketing process and query and reporting subsystems.

much of its power. To support the environment, specialized tools and skills are required.

Marketing tools

Although most statisticians would like to believe that they do the job all by themselves, they actually cannot do the work without some sophisticated software. Statistics problems in the telecommunications arena involve such large volumes of numbers and such complex cases that they cannot be handled with pencil, paper, and a calculator.

- *Spreadsheets.* Not only do spreadsheets make query and reporting simple and friendly, they also allow statisticians to do some of their analytics work. Most spreadsheet packages include the capability to do regression, analysis of variance (ANOVA), factoring, smoothing, and other statistical functions.

- *Statistics packages.* The fact that spreadsheets can do statistics is a nice feature, but they have their limits. Spreadsheets cannot work very well with large numbers of records. This is where the *real* statistical tools come in. The three most popular products on the market in the late 1990s are SPSS, IBM's Intelligent Miner, and SAS. These products include not only the traditional statistical capabilities, but many neoclassical tools such as neural nets and CHAID.

- *Specialized tools.* For marketers with highly specialized needs, there are specialized products that focus on the solution of particular marketing problems. ASA's ModelMAX, for example, is a neural net that can be used for customer list scoring. Because this tool is specialized to support only this function, it is easier to use and more efficient to operate than a generalized tool that must first be "tuned" by the marketer.

Format

While query and reporting tools concentrate on fancy output and pleasing design, analytics software packages, for the most part, are concerned with the math, not the aesthetics, of what they are displaying. Analytics tools output graphs, tables, and miles upon miles of reports, all of which are enough to give anyone a splitting headache. Some statistical analysis tools give some attention to formatting, but format is generally a function of the statistics in use.

Information databases

The information to support analytics includes everything in the query and reporting environment, everything generated by the process management systems, as well as their own unique set of input stores. These unique data stores include:

- *Segment information.* One of the biggest jobs of the analyst is to perform segmentation studies. Once these segments are defined, a good marketing database keeps track of each customer, the segment to which they are assigned, and on what day. By maintaining this history of overlapping segmentation decisions, the analyst produces a rich new store of history data to be mined.

- *Scoring and function information.* Scoring of customers and prospects is another of the analyst's jobs. Although the immediate need for these scores passes relatively quickly, the history of scores should be recorded. The same holds true of customer value functions and other functions created to assist in the marketing process.

- *Call detail records.* The analyst is the most likely person in the organization to make good use of call detail record information. Mining of call detail records can provide a vast new source of potentially useful information.

- *Outside data.* Analysts are also likely to be found working with purchased or leased data from outside organizations. Credit card companies, banks, and professional list brokerage firms all sell raw data about customers, which the analyst can incorporate into the model building process.

Analytic support skills

This area too engenders its own unique blend of skill requirements:

- *Data analysis*. Although some data analysis skills are required to support analytics, the analysts usually have little use for databases. Most analytics can be performed with simple flat file data stores, with most of the files already built for query purposes.

- *Statistical skills*. The single largest limitation on the organization's ability to make use of analytics is not in the tools or databases, but in the skills of the analysts who use them. The approaches and techniques we have talked about here are difficult to learn and use. Many organizations hire only statisticians with Ph.D.s; others look for people with good analytical ability and send them to statistical analysis classes.

Analytics and the marketing process

Only highly trained analysts access the analytics subsystem. Figure 17.2 shows that this system is used almost exclusively as part of the modeling process and is only rarely referenced during prioritization and goal setting and during campaign planning.

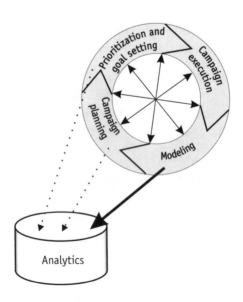

Figure 17.2 Analytics and the marketing process.

Process management

The last, but certainly not least, of the marketing subsystems are those that manage marketing processes. This area is unique in our coverage of marketing databases because, unlike query and reporting and analytics, process management software is not part of data warehousing. Process management means exactly what it says: software that manages human processes (in other words, these are OLTP systems).

Major process management subsystems

Most organizations have some kind of management system in place. Computerizing the process makes managing the various systems much more efficient, effective, and economic. Figure 17.3 shows some of the

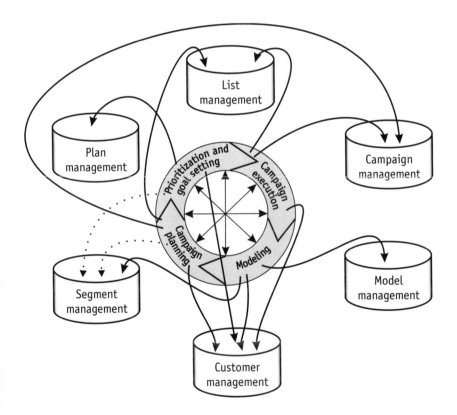

Figure 17.3   Process management and marketing process.

major process management subsystems that a telco may have and their relationship to the overall marketing process.

- *Customer management.* Customer management systems are found in many organizations. Usually sponsored and run by the customer service organization, they hold information about customers and try to function as a centralized clearinghouse for all customer activity.

- *List management.* List management is such a large and complex topic that we have dedicated Chapter 18 to this subject. For now, let's just say that list management is about processing and keeping track of the different sources of customer names, addresses, characteristics, and behaviors.

- *Segment management.* In some organizations, the identification, definition, and changing of segmentation strategies is so fundamental to operations that a separate system is built to manage it. These systems track all of the individuals in all of the different available segments and provide a centralized place for redefining and propagating them.

- *Campaign management.* Campaign management is also a very complex topic and is covered in a separate chapter, Chapter 19. Campaign management actually has two meanings. In the case of direct marketing, campaign management software keeps track of each name on a list and all of the scores and campaigns of which they are a part. This allows direct marketers a better view of the cumulative effect of campaigns. Campaign management also refers to establishment of standards for the execution of both direct and mass marketing campaigns and the measuring of results.

- *Plan and model management.* In some of the most sophisticated organizations, plans and models are managed by software as well.

Process management tools

Because process management systems are OLTP, rather than data warehousing types of applications, there are no tools per se to support them. Instead, there are full-blown applications.

- *Packages.* The marketing industry as a whole is flooded with marketing process management software. It is possible to buy packages that purport to manage the whole marketing process, list management, campaign management, and many other processes. Many of these systems even integrate analytics and query. Unfortunately, there are no packages like this geared specifically to the telecommunications industry and the determination of a "good fit" is extremely difficult. Many organizations end up installing these packages and then working around the shortfalls.

- *Custom-built software.* By far the most common way to get process management software in the telco is to build it. This process can be expensive, but the solutions are naturally well suited to the needs of the telco.

Process management skills

The skills required to support process management are the same skills that support any other business process. Many telecommunications companies function without any process management software and depend solely on the skills of their managers to make the process work. Others lean on the software to augment these skills. In all cases, there are no specialized skills per se involved in this area, except, of course, for generalized marketing and telco expertise.

Process management information stores

Basically, every process that is managed by software will be supported by its own specialized data store(s).

18

List Management

He's making a list, checking it twice,
Gonna find out who's naughty or nice,
Santa Claus is coming to town.
Popular Christmas carol

In Chapter 17 we looked at many of the different support capabilities that a marketing database can provide to the organization. We saw that a sizable contribution to the overall efficiency can often be realized when the marketing database is used to help manage the different marketing processes (i.e., the list, plan, model, customer, campaign, and other forms of management). A detailed exploration of each of the processes would surely provide valuable information about how marketing works in a typical telecommunications firm. Because of the many differences in how individual companies manage these functions and for the sake of space, however, we choose to address only the two most important processes. This chapter covers list management, and in the next chapter we consider campaign management.

List management

We have decided to give coverage to this particular management area for several reasons. First, list management is an extremely complicated and computer systems-dependent process. Most of the other areas can be handled with a human management process. Second, good list management is a very important capability for any telecommunications company to have if they are to take part in direct marketing operations. Third, effective list management can save the telco significant time, money, trouble, and confusion.

List management and prioritization and goal setting

Although list management itself usually plays only a minor role in the prioritization and goal-setting process, there are two areas where it can be crucial in the early phases of a campaign's life.

1. *Existence of a list.* The fact that a particular list is available can actually drive the entire marketing process. For example, say you found out that you could purchase a list of your competitor's unhappy long-distance subscribers. What would you do? Would you create a campaign based on that list? Certainly, you would! Although a perfect list like this is probably a fantasy, situations will arise in which well-focused lists become available and the marketer or sponsor initiates a project based on its existence.

2. *Capabilities evaluation.* List management capabilities come into play during prioritization and goal setting when sponsors and teams struggle with what approach they might take with a campaign. An efficient list management process can inspire people to explore marketing methods that have never been attempted.

List management and modeling (segmentation)

The second place where effective list management can be useful to the marketer is in the area of model development and, specifically, segmentation analysis. Most people associate list management most readily with direct marketing campaigns, but it can be utilized just as effectively in the support of mass marketing operations.

Using internal lists to initiate the segmentation

A list of customer names is absolutely essential to initiate any kind of segmentation activity. Typically, the marketing organization will start by asking for any kind of list with customer names and characteristics that can be extracted from the billing system.

Using external lists to enhance and enrich the modeling process

Most telecommunications companies have a woefully minuscule amount of information about their customers in their billing systems. That is why they turn to other sources to enrich their basic lists. There are hundreds of organizations in the business of compiling lists of names of people and information about them. Some list brokers or list houses make this their primary line of business. Others sell the enhanced lists of customer information from their own files. With an external list, all the marketer has to do is simply "match" it to the internal list. The result is a much richer set of customer information.

Segmentation with and without lists

Some segmentation, by class or category for example, can be done without a list, using instead focus groups, questionnaires, and other kinds of mass market sampling. The best source of information about customers, however, is a good list. Most of the segmentation approaches we discuss in this book are list based.

List management and direct marketing campaigns

List management is the key ingredient of any direct marketing plan development. Any time direct marketing is the medium of choice, the campaign is dependent on the sophistication and efficiency of the list management process. List management gives the marketer new rosters of people to market to. It also provides additional rich sources of information about customers and prospects.

Marketers launch their direct marketing campaigns from individual lists. In this case, the job of list management takes on an interesting additional set of complications. When a company does a lot of direct marketing, they tend to purchase large amounts of external data and pretty soon they have many different lists to manage. Without a system and a capability in place for managing those lists, they would soon be overwhelmed when trying to keep track of which lists were used where and for what.

Major components of the list management function

Given the broad range of situations in which list management plays a critical role, we need to understand its different aspects and how they fit into the overall marketing process. The major components of the list management process follow:

1. *List acquisition:* buying or leasing alternative sources of customer/prospect information;

2. *List construction:* building the list from internal data sources;

3. *List preparation:* making external or internally developed lists useful to the list manager;

4. *List integration:* combining, dividing, and integrating different lists;

5. *List utilization—segmentation and modeling:* performing segmentation analysis;

6. *List utilization—campaign planning and execution:* running direct marketing campaigns;

7. *List usage tracking:* tracking customer contact activity.

List acquisition

List acquisition is usually the first—and least appreciated—part of list management. Many organizations handle the process in an incredibly slipshod and haphazard manner, with a corresponding degree of manageability and accountability. The main reasons why an organization would take on an external list are to start a new list or to enhance an existing one.

A telco can acquire external lists of varying quality and usefulness for both consumer and business prospects.

Sources for consumer lists

Depending on the country where the telco is located and the type of marketing they are planning, a fairly wide assortment of consumer information is usually available for sale:

1. *List houses.* These companies are in the business of compiling lists of names and attributes for consumers. At any given time, tens of thousands of consumer lists are available for purchase through the many facets of the direct marketing industry. (These lists are available for consumers throughout the world to differing degrees of quality.)

2. *Catalog and direct marketing organizations.* These organizations phone consumers directly or send catalogs to them. For the right fee or arrangement, many of these firms are willing to share their lists.

3. *Retailers.* Retail companies can be rich source for list acquisition.

4. *Credit agencies.* In many countries, organizations that track people's credit ratings can sell information about those consumers.

5. *Credit card companies.* More and more consumers are shifting the majority of their purchases to credit cards. This means that credit card companies are keeping vast transactional history records. Access to these records can often be purchased.

6. *Survey companies.* Many companies specialize in conducting surveys, which can build or enhance a telco's list capability.

7. *Consumer research firms.* Related to survey companies, these organizations build extensive consumer profiles that can be purchased by marketers.

8. *Related industries.* An excellent source for consumer names is to exchange lists with another company in a related industry. For example, a paging company and a cellular firm might swap lists.

9. *Magazines and newspapers.* Many magazines and newspapers keep extensive consumer information files and make entire side businesses out of selling these lists.

10. *Government agencies.* Government census data, along with other governmentally funded survey and consumer research information can contribute greatly to the richness of a company's list.

Sources for business (commercial) lists

The telco interested in working with businesses will find similar sources for commercial customer lists. Common sources include list houses, commercial catalog houses, wholesalers and retailers to business, and credit agencies.

Some sources are unique to the business-to-business (B2B) marketer:

1. *Trade organizations.* The organizations that support a particular industry usually compile lists of member companies and vital statistics about them. These lists are often free or very inexpensive.

2. *Chambers of commerce and related organizations.* The traditional bastion of business information, the city chambers of commerce can be extremely useful sources of business statistics and support information.

3. *Government agencies.* Most governments keep incredibly detailed lists of information about companies in their own as well as in other countries. The U.S. Department of Commerce and the equivalent organizations in other countries actually give away rich stores of information about companies, all in the name of promoting trade in particular economic segments.

4. *Magazines and trade publications.* Magazines and trade publications that specialize in a particular area usually make lists of companies and corresponding information available for resale.

Costs for external lists

The costs of lists can vary greatly. In the best case, lists are gratis (such as those from a government agency). Another way to obtain a list is via an exchange. You may swap one list for another, exchange a list for advertising, or offer some other trade.

Most often, lists are either sold or leased for a fee. Lists are usually costed out on a per-name basis. Large consumer files may go for fractions of a cent per name, whereas small, highly focused lists can cost as much as several dollars per name. The more targeted and accurate the list, the higher the cost per name will be.

Quality and utility of external lists

Just as costs for lists vary, so does quality. Lists can be new and accurate, or old and "stale." A stale list is one with a lot of out-of-date addresses and inaccurate customer information. Aside from being inaccurate, lists can also be "dirty." A dirty list requires much formatting and preparation before it can be used. Obviously, ascertaining the quality of a list is an important part of list acquisition.

Managing external list utilization

In the final analysis, the acquisition of lists must be managed like any other process. Too often, organizations will allow anyone to purchase a list without clear list acquisition guidelines. The result is an incredible amount of waste for the organization.

List construction

Not every telco acquires external lists but, without a doubt, every telco gets involved in the construction of lists. Luckily, building a base list is usually simplified by the nature of the telco information systems architecture. Because most telcos keep the vast majority of their customer information in the billing system, creating a base customer list usually involves little more than extracting and formatting information.

The fact that the billing system is the primary source of list data does not mean that the process is trivial, however. List construction requires that the marketers and the IT departments identify all of the appropriate sources of data, guarantee their accuracy, and create a technique for continuously refreshing and updating that information.

Synchronization issues

Making a copy of the billing system and then building the first list may seem simple enough, but someone needs to think about how this process of extracting from the billing system and updating it will be managed in the long run. Although a one-time extraction can be useful, this technique cannot be used continuously because the base list that the marketer uses must have "memory." That is, it must remember who was, is, and will be a customer, regardless of the status assigned by the billing system. For example, if a customer churns much (drops and reinitiates service), the simple billing extract approach will not reflect this because the customer keeps appearing and disappearing from the list. In addition, the billing system will not keep track of everything that the list will hold. As the marketer enhances the list, those enhancements need to be saved.

When building a base list, therefore, a regular billing system update mechanism needs to be included that allows the base list to maintain its integrity, while simply being updated by changes in the billing system.

List preparation

Once the list has been purchased or constructed, the marketer needs to prepare it and make it usable. To be useful, a list must meet certain standards regarding quality (the integrity of different fields) and homogeneity (a list must be "mergeable" and compatible with other lists). Because of this, the following major list preparation disciplines have been developed over the years.

Data field integrity

To make a list useful, the first issue of import is to check for content and domain integrity, making sure that each of the data fields holds whatever information it is supposed to.

- *Content integrity.* This term describes the quality of the data held in each field. Many times, lists will have fields that hold little or none of the information they are supposed to hold. For example, a numeric field might hold "xxx" or any of a number of other field content problems.

- *Domain integrity.* Another difficult problem to address is domain integrity. This refers to whether or not the values in a field are of the correct type. For example, an age column should hold values between 1 and 100; the field for gender should hold an M or F, no other letters.

- *Context integrity.* The most critical integrity issue to be addressed is the accuracy of the data. For example, is the phone number or address listed for the person correct?

In all cases, integrity management is handled either by a marketer who reviews the list by hand or by a program written to provide those kinds of checks.

Name and address hygiene

The names and addresses on a list are incredibly crucial to its usefulness. Maintenance of the validity of these values is, therefore, a specialized discipline within list management. Included in name and address hygiene are these elements:

- *Name standardization.* The big problem with name files, is that there are so many different ways to spell a person's name. "J. Smith," "John Smith," and "Johnny Smyth" might all be the same person—or maybe not. When two lists are compared (an internal and an external list, for example), the name should provide the means to match up the lists. To provide this matching, there are special name matching algorithms, many of which incorporate *soundex,* a technique that translates the spelling of names into their mnemonic base sounds then matches the names accordingly.

- *Address standardization.* Almost as challenging as the matching of different names is the matching of disparate addresses. Addresses can have the same level of inaccuracy and inconsistency as names can and similar algorithms are used to match them up.

- *Special name and address hygiene software.* The cost and problems of name and address hygiene can be so overwhelming for many organizations that they either purchase specialized name and address software or pay an external firm to do the "cleansing."

Address and phone number accuracy

While name and address hygiene are important, making sure that the addresses and phone numbers on file are correct is even more important. After all, they are your contact points. There are really two approaches to managing this accuracy problem:

1. *Feedback mechanisms.* The most accurate information comes from the telco's own organizational structure. Updated name, address, and phone information in the billing system should be forwarded to the marketing database as quickly as possible. Additionally, any time a direct marketing activity reveals an inaccuracy (returned mail or wrong phone number), that information should be forwarded to the billing system as well.

2. *U.S. NCOA.* The National Change of Address (NCOA) group publishes lists of address changes in the United States and many other countries. Such lists can be used to keep addresses fresh and current at all times.

Customer/prospect keys

The last item that many organizations incorporate into their base lists is a list *key,* a customer number or prospect ID. This basic key structure allows the marketer to attach a stable, unchanging identifier to each customer and prospect with which the organization keeps track of people's history as they change address, name, phone service, and anything else.

List integration

After the list has been cleaned and made ready for use, the next step usually involves integrating it with other lists. This process can be done in any number of ways.

Match/merge

Match/merge is the most common way to join two lists. The two lists are sorted in alphabetic order by the consumer or company name and then matched. The contents of both records are merged into one larger, more robust record.

Merge/purge

Another way to compare lists is to *merge/purge*. The lists are matched, and the record with the better or more accurate data is selected as the new record.

List leasing and blind merges

One of the ways in which list houses and catalog organizations protect their lists is to lease the lists, rather than sell them. In a list *sale*, the customer acquires a copy of the entire list. In a list *lease*, the list owner agrees to allow the lessee to mail a piece to their list (blind mailings) or to run a blind match/merge.

Blind mailing

When an organization contracts to send a direct marketing piece to a list of customers owned by some other organization, the process is called a *blind mailing*. In this case, the telco prepares the mailing pieces and sends them to the sender, who then affixes the labels and mails them. This approach is generally used to try and sell something to the customers. Another reason to use a blind mailing is to "leach" the other's list. In that case, the mailing will ask the consumers to fill out a form or send in for some gift. This technique allows the lessee to gain ownership of the names on the lessor's list.

Blind match/merge

In a blind match/merge, the organization attempts to gain additional information about their own customers through the use of someone else's list. The purchasing company provides a list of all of its customers, including name and address, to the leasing firm. The list lessor then performs a match/merge between its list and the telco's list. For those customers who show up as a match, the attributes associated with the lessor's file are given to the telco. In this way, the lessee is able to gain more attribute information, while the lessor is protecting their list.

List purchase

In some cases, the marketing organization will simply buy, swap, or attain a list for free. In these cases, the newly acquired list will be prepared, matched and merged, and added to the inventory of the organization's lists.

List utilization: segmentation studies and modeling

Once the lists have been obtained, prepared, and integrated, the marketer is ready to begin segmentation and modeling activities. (See the earlier chapters on modeling and segmentation for more information about these processes.)

List utilization: direct marketing sourcing

Ultimately, the lists obtained by the company will be used to help drive the direct marketing planning and execution process. (Please refer to the earlier chapters about campaign planning, execution, and scoring, and see the next chapter about campaign management for more information about how list management can be used to enhance campaign management.)

List usage tracking

The value of a large inventory of lists within the organization can be greatly enhanced by simply keeping track of where all of the different lists came from and how they were used. This is particularly important in two situations:

1. *Stale lists.* The number one joy of any direct marketer is a list of good prospect information. Their number one bane is that same list after it was used to mail dozens of campaigns to the same group of prospects. People on those lists can quickly become sick and tired of getting direct mail pieces and receiving phone calls from the same organization over and over again.

   When the same list is used too many times by the same organization it becomes stale. There is no cure for a stale list, except to stop using it. Many organizations resort to *list freshening,* which involves adding new customers to the list so that the percentage of people receiving the same old message is reduced.

2. *Controlling list expenses.* The second reason that keeping track of lists is so crucial is because many times, in larger organizations, the expenses involved in list management can get out of control. One organization I worked with discovered that four different areas of the company were purchasing the same list, spending three times the amount of money necessary.

19

Campaign Management

Reach Out and Touch Someone
AT&T long-distance advertising campaign

Connections so clear "You can hear a pin drop."
Sprint long-distance campaign

Campaign management is one of the most important—but also most confusing—of the marketing processes. The list management process we discussed in the previous chapter is nearly impossible to manage without computer systems. Campaign management, however, is usually so loosely defined and executed that it is hardly ever managed with the help of computer systems. Unfortunately, this can be a mistake, because campaigns are the lifeblood, the basic foundation, of what the marketing organization does. Without a good campaign management process in place, many of the best efforts of the marketing organization can be wasted.

What do we mean by campaign management?

Many people and many software firms claim to have the answer to your campaign management problems. Unfortunately, as we will discover, what these vendors consider to be campaign management and what it means in the context of a telecommunications marketing operation are usually significantly different. We, therefore, need to develop a definition for exactly what we mean by the term *campaign management*.

The marketing process and the role of campaigns

Let's very quickly review what we know about campaigns. Within the context of the overall marketing process, campaigns are basically born out of the modeling process. Sponsors identify objectives and assign teams to investigate options in the prioritization and goal-setting step. At this time the sponsor creates a "project" that describes the objectives to be addressed, some of the alternative approaches, and any constraints the sponsor thinks appropriate.

After this, the team is formed, goals are set, the identity of the project is established, and then the team moves into the modeling phase. During modeling, teams develop alternative models for execution (identifying market segments, messages, media, time frame, and prices) and recommend the best ones for execution. Usually, each project initiated by a sponsor will lead to the development of several alternative models, each attempting to best meet the project goals and constraints.

At the end of the modeling phase, the sponsor selects the best or preferred model for the construction of a particular campaign. The newly initiated campaign will then be put through the campaign planning process for detail development, testing, and finalization.

Finally, the campaign itself is executed. A review of the campaign life cycle is shown in Figure 19.1.

So, the process of campaign management can be said to start at the point when different marketing projects are born, continue through model development and selection, and proceed through the actual creation, development, and execution of the campaign itself.

A comprehensive campaign management system, therefore, if it is to be truly effective, needs to keep track of and manage the entire marketing

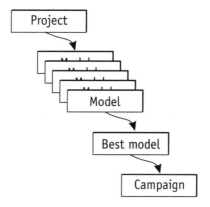

Figure 19.1 Life cycle of a campaign.

process, including aspects of goal setting, modeling, planning, execution, feedback, and measurement.

Campaign management software

One of the most exacerbating things about the marketing process, and campaign management in particular, is looking for a campaign management software package that can manage the whole process. Unfortunately, what can be said about the majority of campaign management software on the market today is as follows:

1. They were written with a certain industry in mind (usually direct marketing or retail, sometimes banking, but *never* telecommunications).

2. They assume a certain size of organization (different sized companies selling different products market them differently).

3. They assume a certain organizational structure (a structure that telcos never have).

4. They assume that the organization does only direct marketing (a situation which is almost never true in the telecommunications case).

5. They focus on direct marketing only.

6. They make no allowance for integration of sales, channels, advertising, customer service, or any other nondirect marketing measurement into the formula.

7. They combine many different aspects of marketing management, including campaign, plan, modeling, list, and other forms of management, into a unique and far from comprehensive, but usually haphazard, organization of partial management capabilities.

Comparing campaign management software packages

If you do decide to investigate the purchase of campaign management software packages, then we recommended that you consider the following questions/issues as a part of that investigation:

- *How does the product manage direct versus mass marketing campaigns?* Examine how the software allows you to include mass marketing campaigns and promotional activities (the majority do not).

- *How much of the marketing management process does the software handle, and how?* Closely review with the vendor exactly what marketing management functions the software actually manages and challenge them to explain the workings in detail.

- *How does the software manage teams, models, and campaigns?* Make a list of the different marketing management processes, components, and participants (sponsors, objectives, teams, plans, models, segments, product lines, prices, profits, campaigns, scores, functions, feedback) and apply it to your understanding of the software. How does the software record and manage the existence of and interrelationships among these parties, processes, and components?

- *How well do the processes instituted by the software fit with the way your marketing process works?* The real test of the fit of a software package is in understanding how well it can help your organization manage the process and itself better. Ask the vendor to walk through the marketing process and explain how the different functions will be enhanced with the software.

Campaign management: best practices

Some people can spend an infinity of time arguing about what campaign management is, but we prefer to take a more pragmatic view. Let's consider what the campaign management process would be like in the ideal situation, and then see how we might institute a system to help make that process a reality.

Campaign management goals

The first issue that we must wrestle with is to figure out what the goal for the campaign management process should be. There are many views on this and we will outline some of them.

Manage campaign planning activities

One of the most obvious roles a campaign management process can fulfill is managing the planning of individual campaigns. Of course, the amount of work involved in the planning of a campaign will vary greatly based on the media used. Direct marketing, with its intensive list management burden, is often a logical target for this kind of activity. However, trying to impose a formal process or software management on the execution of mass marketing and promotional activities is likely to prove to be a frustrating waste of everyone's time.

Manage campaign execution

Another place in which the management of campaign activity can be exceedingly helpful is in campaign execution. Many organizations could use a facility that allows them to track the progress of each campaign so that they can modify or cancel the campaign based on the feedback received.

For example, one organization I know of developed a campaign that included coordinated execution of a series of television and newspaper advertisements over a period of 4 weeks to promote the introduction of a new cellular service in the area. The offer was so good, and the market so receptive, that the telco found itself quickly overwhelmed in requests for the new service. The signup rate was so high that, by the fourth week, they were turning down and antagonizing more people than they were signing up. What this company needed was a warning system, built into their marketing process, in which the marketer tracked the early results of the advertising. When the extremely high activation rate was noted during the first or second week, the marketer could have cancelled, postponed, or changed the ads scheduled for the following 2 weeks and prevented the public relations and operational nightmare that was created.

In another situation, the marketing department might be committing to the execution of an aggressive telephone campaign. Let's say the campaign, which is run for several months, involves placing phone calls to thousands of people (at a very high cost per call!). At times, however, the response of the public to a phone solicitation program is not what marketing expects. In fact, the marketer may find that, after only 2 or 3 days of calling, the expected response rate is not being achieved. In this case, the marketer will want to cancel the campaign and save the marketing expense that will clearly not yield a decent return on investment.

Post-campaign analysis

Most marketing organizations in telecommunications use only the crudest and most simplistic of campaign impact measuring techniques. Typically a marketer will run a campaign and then check the weekly numbers posted for activations. If the number of activations goes up, the campaign is considered a success. This technique, while indicative of success in a general sense, is far from an accurate or reliable measure. This simplistic feedback mechanism does not help the marketer sort results based on the cumulative effects of other campaigns that might be running (for example, when advertising, customer service, and sales all run proactive acquisition campaigns at the same time).

(Many marketing organizations allow this flaw in the post-campaign analysis to exist because it enables all organizations to "double count" and take credit for the same increase.)

Measuring the quality of the customers acquired by the campaign

Being able to separate out which campaign yielded what kinds of results in terms of raw activation numbers is significant, but there is an even more important issue to consider. The measurement of campaign results done by most marketers does not involve any consideration of the quality of the new customers acquired. In this world, acquiring new customers is all that counts. Whether they are bad credit risks or low-volume, high-cost users is not addressed.

Measuring the cumulative impact of all campaigns

Finally, what we would really like to have is the ability to track not only how effective individual campaigns are, but to actually determine what cumulative effect all of the campaigns that a person has been subjected to has had on their decision to join and stay with the company.

The ideal campaign management system

Ideally, a good campaign management system would be one that takes all of these things into account. It would help manage the following for all types of campaigns and many different levels of detail: (1) project initiation and goal setting, (2) model development, (3) campaign planning, (4) campaign execution, and (5) campaign effectiveness measurement.

Campaign management: media idiosyncrasies

It is one thing to talk about campaign management from a theoretical perspective, and another to make it work in the real world. There are two constraints to making this idealized view of campaign management a reality. The first has to do with the differences between the media and the way they must be managed, and the second is based on the technology and data available to do the tracking.

We consider first the media differences.

Direct marketing

For anyone wishing to do a good job of keeping track of customers, messages, and impacts, the world of direct marketing is the perfect environment to be in. In the direct marketing universe, the marketer has almost complete control of the entire marketing process. The decision to send the message to an individual is made based on information about that person that is already available. Each individual that will be approached with a message is listed and identified by a large number of characteristics.

Because of the well-controlled nature of direct marketing, the marketer can time the execution of the campaign in a way that allows for monitoring of the effectiveness of the campaign as it progresses, enabling "in-stream" adjustments to the campaign. Each time a message is sent to that customer, that fact can be recorded. Therefore, each time the customer responds, buys something, or increases or decreases use of a product, we can tie that change in behavior directly back to the campaigns.

Indeed, the direct marketing world defines the template for how we would like to see all campaigns managed. With the right software process in place, the direct marketer can manage campaigns at the planning, execution, and post-analysis phases with amazing precision and accuracy. At the same time, the marketer can also evaluate the impact that cumulative campaigns will have (as long as all of those campaigns are subjected to the same direct marketing campaign management discipline).

Mass marketing and promotion: management challenges

Direct marketing is actually only a part (and sometimes a small part) of the activities that a telco will participate in on the marketing front. This, of course, creates a problem for the marketer who wants to be able to manage and measure the progress of all campaigns.

The world of direct marketing is rich in two things that the manager of the marketing process needs to know about:

1. The direct marketing process records every single major activity in which a customer is involved. That is because in direct marketing everything derives from the list and every time a list is accessed, scored, or evaluated, the activity is recorded on the customer's master record. This means that, if anyone wants to track the progress or the history of contacts or decisions made about a customer, all they need to do is pull up that file and review it.

2. The direct marketing process keeps track of individuals. Because of this, the marketer can keep track of individual people. Unfortunately, that is not true of mass marketing. In the mass marketing and promotional worlds, customers are analyzed and dealt with as groups or blocks of people. The mass marketer does not deal with individuals, so that kind of measurement would seem to be impossible. Mass marketers have only a general idea about which individuals received a message and no clue as to how many messages. Because of this, the mass/promotional mar-

keter's only feedback comes from one of two sources: (a) *A buying/consumption event:* The mass marketer knows that they succeeded only by inference. In other words, say, the company has 100,000 customers today and the mass marketer runs an ad. If the number of customers goes up to 125,000, then the marketer will infer that the ad generated the business. (b) *Indirectly through surveys, focus groups, and so on:* The only other way for the mass marketer to find out what effect their activities had is to question a random sample of typical consumers. Again, they infer that their results were consistent with those findings across the entire population.

Making campaign management work for all media

The basic differences between direct and mass/promotion marketing from the management perspective are significant. The direct marketing approach is rich in detail and specific data. The mass/promotion approaches involve very little data and make most of their conclusions based on inference.

The challenge for the marketer who desires to do a good job of managing the overall marketing process, and campaigns in particular, is to come up with a way to blend these two forms of tracking and levels of detail into the same system.

Identifying common denominators

The first step would be to figure out how to get all reporting to be consistent across the three media (direct, mass, and promotion). Luckily, we can turn to several common dimensions:

- *Segments.* The first common denominator that unites the different marketing media is the segments used to describe different groups of customers. There is no reason why we cannot define different segmentation schemes that all of them share so as to provide the organization with a common baseline for tracking, managing, and measuring the effects of campaigns.

- *Campaigns.* Although segments can provide us with a common way of looking at customers, the fact that the three media execute different kinds of campaigns does not mean that those campaigns cannot be registered and managed from a common perspective. Although a direct marketing campaign is different from a booth at a county fair promotional event, both are really just different forms of campaign execution. Each involves a message, media, market (segment), margin (price), and specific objective (wallet share, acquisition, retention). Therefore, each can be recorded and compared for its effectiveness according to the same criteria.

- *Customer transactions.* Finally, the glue that binds these three forms of marketing together is the customers and their individual transactions. Promotion and mass marketing may use inference to measure their results, but that inference is based on detailed transaction records. Therefore, the results of all campaigns can be measured against this same common set of criteria.

Multiple-classifications requirements

To track campaigns for all media, all you really need is a mechanism that tracks the common indicator variables throughout the development and execution of the campaigns. This means that each area of marketing activity (the marketers working with different sponsors, teams, and media) simply needs to be cognizant of the commonly agreed on structures and make use of them as an adjunct to the structures they already use. In other words:

- All marketers need to understand and agree on the meaning of the terms *project, sponsor, team, objective, model,* and *campaign.*

- A common set of campaign identifiers needs to be established and utilized by all marketers. These identifiers can then be used to tag each campaign event as it progresses, is executed, and is analyzed.

- A common set of segmentation identifiers needs to be established and utilized. These identifiers can be used to "double tag" each customer or marketing event.

The information generated by each of the three types of marketing effort can be pulled together and analyzed based on the common identifiers.

Campaign management in action

How does this campaign tagging activity work? The scenario would go something like this. A common system would be established to manage campaign activity across the organization. Each time a sponsor decided to have a campaign run, he or she would be required to "register" the campaign in the campaign management system. This registration process would require that the marketer report these items of information:

- Who the sponsor of the project is;

- What the objectives for the project are;

- What models were developed to justify the campaign;

- What model was chosen as the basis for the campaign;

- Specific identification and registration of the message, media, market (segment), margin (price), and time frame;

- For direct marketing campaigns, tracking would continue by lists, scorings, mailings, and so on (all events would be tagged with the registered campaign ID);

- For a mass marketing or promotional campaign, the system would record information about advertisement schedules, newspapers, and television stations, price points, statistics about the demographic segments penetrated by the advertising events, the costs, and so on;

- After the campaigns run, the collected information could be used to analyze the net effect of all campaigns on the customer base.

Holistic analysis

A system like this will require the construction of a very special kind of marketing database environment, one that allows for tracking, tagging, and reporting on marketing activities across a variety of dimensions and time frames. In the following and final chapter of the book, we look at the construction of a marketing database environment from the physical perspective and see how such a system can be made possible.

20

Marketing Database: Architecture

We shape our (systems) buildings, thereafter, they shape us.

Winston Churchill (paraphrased)

The physician can bury his mistakes, but an architect can only advise his clients to plant vines.

Frank Lloyd Wright

Less is more.

Mies Van Der Rohe

In this last chapter, we address the nuts and bolts, the mechanics, and the physical realities of building a marketing database system that can support the marketing processes that we have been evaluating. The reason for waiting until the end is so that we have a baseline understanding for how complicated the marketing process is for a telecommunications organization and so that we can appreciate how important each of the functions and computer-supported processes is to its overall execution. Now we can discuss how this myriad assortment of processes and analytical needs can be supplemented, supported, and structured by creating a comprehensive, well-designed, and *well-architected* marketing database system.

The custody and chaos of marketing database systems

The marketing process is, unfortunately, not managed very well in most of the telecommunications world. Using computers to help manage the process often adds to the confusion created by the many different sponsoring organizations, media approaches, team memberships, and competing outsourcing providers. Two of the major reasons for this are the chaos and custody issues.

Chaos in marketing

The process we have been describing may seem structured and organized, but that is only a faint reflection of the reality of marketing. The rate of change, the pace of decision making, and the conflict between different business units contribute to an environment that is far from structured.

Custody

Attaining funding and designing a comprehensive marketing database system that takes all media, projects, organizations, and phases into account is an extremely expensive and politically difficult process. The causes generally revolve around ownership issues.

First, there is the problem that all organizations participate in the marketing process, but none claims total ownership. Second, because the level of participation of each group is limited to its own specific objectives, funding comes with the same limitations. No one is interested in investing in a comprehensive marketing database solution, only in a part of it. Third, different kinds of media require a significantly different kind of computer support system. Direct marketing support systems tend to be very different from those used for mass market support.

Function before architecture

This all leads up to the one concept that we started with in the introduction to this book. That is, you cannot begin to develop the architecture and design for your marketing database system until you have clearly defined the parts of the marketing process you will address with what technologies in support of which media. I hope that you agree with this statement because, sadly, it is the exception, not the rule.

Build it first, use it later

In the majority of cases (perhaps more than 95%) people use a basic two-step thinking process in the decision to build a marketing database system. Step one, build the system. Step two, figure out how to use it. Of course, the marketing database area is much too complex and much too unique to the telecommunications industry for this approach to work. The first step has to be to develop a basic inventory of functions that the marketing database is to support.

Query and reporting systems

You need to define the system's query and reporting capabilities in terms of the data and the tools to be provided.

Consider these types of information:

- *Customer information.* Identifies the nature and extent of customer information that will be made available.

- *Product information.* Provides analysts with information about the activity and costs associated with different product lines.

- *Call detail information.* Allows viewing of individual call transaction information.

- *Billing information.* Provides a gateway into the billing system.

- *Other data sources.* Other possible information sources that can be made available.

The following types of tools can be provided:

- *Ad hoc query tools.* Include spreadsheet interfaces and formal query products.

- *Report writing tools.* Provide sophisticated formatting and summarizing capabilities.

- *OLAP capabilities.* These are used when advanced multidimensional analysis is appropriate.

Analytics

We also need to define what analytics functionality will be required. The various types are as follows:

- *Segmentation.* The different segmentation schemes that support marketing efforts.

- *Functions.* Calculation, computation, and assignation of customer value functions, lifetime value functions, and others.

- *Scores.* Assign the different types of scoring algorithms to the customer/prospect population.

- *Models.* Using analytics for other kinds of calculation and exploration.

The analytics tools that are available to assist the marketer in these calculations include the following:

- *Spreadsheets.* For basic analytical work.

- *Statistical analysis tools.* For more sophisticated factor, regression, and other classical statistical analysis.

- *Data mining tools.* To support advanced analysis using neural networks and other neoclassical approaches.

Process management

As we have discovered, there are many processes that need to be managed in the marketing operation. To design a process management system, we need to identify the specific processes to be managed by the database. These include, but are not limited to, the following:

- *List management (direct marketing).* This is the most common area for database support. Computer systems are used to manage all of the direct marketer's lists.

- *Campaign management (direct, mass, promotion, and so on).* The second most common area for management software is in measuring and controlling the running of campaigns.

- *Plan management and others.* Many other subsets of the marketing process can be managed as well.

Once we have defined the processes, we must decide which media are to be managed, which can include the following:

- *Mass media.* Advertising on television, in print, and other media. (The toughest types of campaigns to manage.)

- *Direct media.* Telephone and mail advertising.

- *Promotions.* The easiest to keep track of but the least often managed.

Next, we must ascertain which of the four major steps in the process will be included in the scope of the management process:

- *Prioritization and goal setting.* The process of identifying sponsors, building teams, and setting goals.

- *Modeling.* The process of identifying targets for marketing activities and setting basic directions.

- *Campaign planning.* The process of putting together specific marketing campaign plan details.

- *Campaign execution.* The actual execution of individual campaigns.

Organizational scope

Aside from these mechanical and process-related issues, we must figure out which organizations will use the system and how it will be paid for. This is the most important and the most difficult job of all. After that, the builder of a telecommunications marketing database needs to review the list and develop a description of the database based on those perspectives. Assuming that all political and funding issues have been resolved, we can then begin the real work of architecture development.

Marketing database architecture: overview

Keeping in mind the functionality we expect the marketing database to support, we are ready to put together the physical arrangement. Although there are a few variations, the ideal structure to support this kind of system is known as a *data warehousing architecture*.

Data warehousing

Because the basic functionality of a marketing database involves extracting information from legacy systems, storing it in a database, and making it accessible for marketers, we find that the typical data warehousing approach meets the requirements most effectively (see Figure 20.1).

A typical data warehousing environment has three component parts (acquisition, storage, and access) with corresponding functionality. The *acquisition* component handles identification, extraction, and preparation of data from other sources to make it useful to the marketing database system. The *storage* component addresses the issues of organization

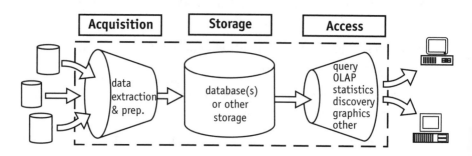

Figure 20.1 Marketing database architecture.

and storage of data for use. The *access* component is concerned with the installation and use of query, reporting, and analytics tools that allow the marketer to work with the database. On top of these functionalities rest whatever process management capabilities that are to be built into the system.

Let's examine each of the components.

Data acquisition

Building a marketing database is very similar to constructing a data warehouse and much can be learned by drawing on experiences we have gained from the latter.

The data warehousing world has found that there is one rule that holds true almost every time when it comes to costing the system. It turns out that the data acquisition area invariably is the most expensive part of these systems. The design, development, and execution of the data acquisition aspects of a marketing database system will generally use between 50% and 80% of a telco's marketing database construction budget. Let's see if we can figure out why data acquisition is such an expensive part of the process.

Data acquisition has three main characteristics that we should be aware of:

1. *Data sourcing.* Where the data will come from for the system.

2. *Data processing.* What needs to be done with the data to make it usable.

3. *Data mechanism.* The mechanical processes used to acquire the data.

We will take some time now to explore each of these in greater detail.

Data sources

For the typical telco, the problems are greatly simplified by the nature of telecommunications systems. Two major systems usually provide the bulk of the data needed for analysis (billing and external sources), and only a handful of other sources.

Billing systems

The billing system is the main source of all information about customer activity for the majority of telecommunications firms. It usually contains the following information about each customer:

- Identity (name, address, phone, and so on);

- Product selections (products customers subscribe to);

- Utilization history (how much of each product they buy);

- Credit and payment history (how much the spend and how well they pay);

- Other background information if applicable.

Often, the billing system turns out to be the *only* source of information about customers. A reasonable and useful extract from the billing system is, therefore, the first and most important step in any data acquisition exercise.

External sources

The other common source of data to be managed by the telco is external. Data purchased from list brokers, magazines, catalog houses, credit card companies, and dozens of other sources provide the telco marketer with a rich assortment of prospect information. Of course, because external data comes from outside the firm, much work needs to be done to make it usable.

Call detail records

For some companies, the analysis of customer utilization information at a much more detailed level makes sense. In these cases, a call detail record repository, created by pulling extracts directly from the switching systems and feeding them into a database, is the favored approach. Call detail record systems are extremely large and have very limited utility, however. They are therefore built only if there is some proof that the information will be of adequate value.

Other internal sources

The larger a telco gets, the more likely it is that other subsystems exist that can provide valuable information. These systems may include the credit system, accounting systems (for product profitability data), and the customer service system (for specific customer contact information).

In all cases, alternative sources must be prepared just as for externally sourced data.

Data acquisition processing

The next problem arises when we try to make that data useful. In general, this process involves two steps: extraction and preparation.

Data extraction

Data extraction is the process of getting the data out of whatever system houses it. In the case of external data, this is not an issue because the provider of the data will already have extracted it.

For billing system data and other internal sources, the process of extraction can pose a number of problems. Putting together an extraction strategy includes several considerations:

1. *Extraction scheduling.* Production systems usually run at full capacity for most of the day. Systems like these are generally not too amenable to the extra stress that an extraction process will put on their processing capabilities. Therefore, finding the right time and place for the extraction can be a key issue.

2. *Synchronization challenges.* Just because we do one extraction from a system does not mean that we are done with it. If our marketing database system is to be useful and timely, then extractions will have to be made periodically and the extracted changes will have to be integrated and synchronized within the marketing database.

Data preparation

After the extraction challenges have been met, the marketing database developer must deal with the problem of preparing the data for use. In fact, there are three important aspects to the data preparation process known as the *three F's: find* it, *fix* it, and *fit* it. Because these three jobs are so crucial to the successful execution of the data acquisition process, we will take some time to review each.

Find it It may be difficult to believe, but the number one expense in data acquisition accrues while people simply try to figure out where the data they are looking for is located. Unfortunately, legacy systems, such as 10-year-old billing systems, do not usually come with a well-documented, easy-to-use data dictionary. These systems are older, not well maintained, and full of data fields with odd names and unclear applications. The only way to make sense out of these systems is to dump their contents and go through a painstaking field-by-field reconciliation of supporting databases with the programs, screens, and reports that the system produces. This process, obviously, can take a long time and be difficult to support.

Fix it Once the data is found, the next problem is to make it useful. Data may be stored in incorrect formats, make use of codification schemes that are invalid, or be erroneously modified by programs. (See Chapter 18 on list management for more information about these problems.) Generally, a field-by-field cleaning, modification, and standardization process must be exercised to make it work.

Fit it As if the challenges of finding the right data and making it useful were not troublesome enough, the marketing database developer has an additional problem. Joining data from two or more different sources always creates difficulties. Alternative data sources hardly ever use the same identifiers to keep track of individuals or use the same periods to measure events. (Again, see Chapter 18 for more information.) The last step of data preparation involves establishing common identifying keys and time frames for all data so that they can be joined with accuracy.

Data acquisition options

Our final area of concern is the actual mechanisms used to perform the work. The mechanics of data acquisition can be approached in a number of ways and may entail the use of several extraction and processing techniques. We review each of them here.

Software packages

Because the extraction and preparation of data for the marketing database is such a big and costly operation, many companies have tried to minimize this cost with special data acquisition software. These packages are designed to alleviate the complexities of data acquisition with menus and fill-in-the-blank screens that ease the process. In theory, these products allow the marketer or IT department to organize the process in a systematic and software-controlled manner. Unfortunately, the use of software packages has had mixed results, at best. The complexities of real-world billing systems make the use of such products viable in only the most limited situations.

Utility programs

One way in which organizations attempt to ameliorate the cost and hassle of data acquisition is to purchase data acquisition utility programs or to use extraction programs built into existing systems. Most billing systems, for example, have a data extraction utility built into their product suite. These types of programs, although limited in scope, are actually extremely useful in easing the burden typically associated with acquisition operations.

Customized data extraction code

Despite the best efforts, the majority of the data acquisition work in support of marketing database development is, unfortunately, accomplished with custom written programs. These programs, usually written in Cobol, C, Assembler, or other common program language, are created to do all the extraction, preparation, matching, and merging work that needs to be done.

Data storage

Data acquisition may be the first step in physically loading the system, but not until the architect has designed the data storage portion. After all, if you do not know what data you need, you cannot extract and prepare that data. Because the development of the data storage area is so critical and complicated, data storage architects have refined levels of design work:

1. *Conceptual level:* the high-level definition of acquisition, storage, and data use;

2. *Logical level:* detailed arrangement of major data collections;

3. *Physical level:* the actual physical organization of data.

This is probably a good place to reinforce the statement made earlier that you need an inventory of the functionalities that you expect the marketing database to support before you piece the architectural environment together. Tragically, many organizations attempt to construct conceptual, logical, and physical data models without any idea about what functions they will ultimately need to support.

When you combine a good functional inventory with a preestablished set of data models you can greatly increase the speed and efficiency of system development.

Conceptual model

The conceptual model describes the first and most basic level of documentation and understanding of the data storage elements of a marketing database environment. It tells the architect which data populations will be included in the data storage area, where the data will come from, and how it will be organized (see Figure 20.2).

The conceptual model of Figure 20.2 shows us the four major "data universes" that make up the construction of the telco marketing database world:

1. *Customer:* information about the customers themselves;

2. *Customer transactions:* information about customer spending patterns, utilization behavior, and other transactions of interest;

3. *Product:* information about the availability, profitability, and use of different products;

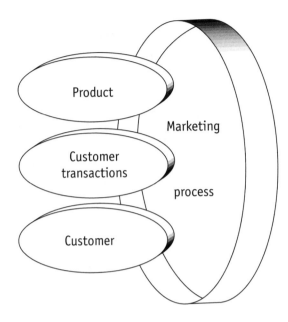

Figure 20.2 Conceptual model for a telco marketing database.

4. *Marketing processes:* information captured as part of the marketing process itself.

This is where information is collected and managed within the marketing database world. The nature of this conceptual model changes depending on the functional inventory of the system. For example, the marketing process universe can be eliminated if no marketing process management will be done by the system.

Data sources

Once the conceptual model is developed, the data modeler determines how each of the areas will become populated (see Figure 20.3). At a minimum, call detail, billing, and marketing process control systems populate typical marketing database environments. Customer data is usually enhanced by externally provided data.

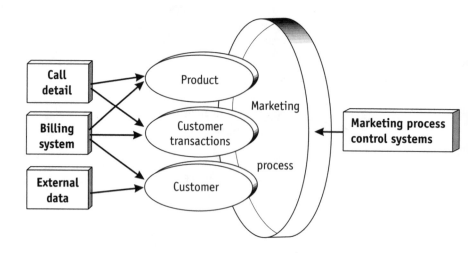

Figure 20.3 Data sources for the conceptual model universes.

Logical model

The conceptual model provides a general idea of what is needed to drive the database, but the modeler will need more detailed information about specific data sets before any serious design work can be done. Logical models are diagrams and documents that describe the major collections of data to be included in the database and their relationships to each other.

Customer model

The one model that we know must absolutely be included in any marketing database is the customer model (see Figure 20.4). A high-level logical customer model will typically include:

- Core customer information (name, address, and so on);
- Characteristics (demographic, psychographic, and other descriptive information);
- Behavior (what customers are buying and the utilization patterns);
- Measures (segments, functions, and scores assigned to the customer).

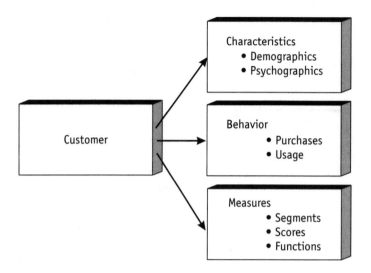

Figure 20.4   Logical customer data model.

Marketing process model

Assuming that the system participates in managing the marketing process, we will also have marketing process information. As we can see from the model in Figure 20.5, the marketing processes will create information about campaigns, segments, models, scores, and functions, all of which will be captured and managed by the system.

Physical model

Once the logical models are completed, the systems developer is ready to create the physical models of the actual databases. Physical models look similar to logical models, but are much more detailed. They vary based on the physical and organizational constraints.

Data access characteristics

Because a large portion of this book is dedicated to the data access characteristics of a marketing database, we do not have to go into it again here. We have provided an inventory of applications, tools, and

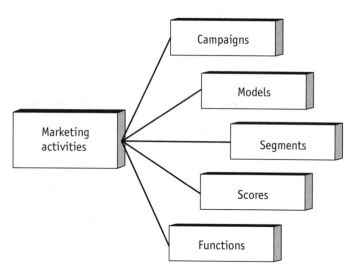

Figure 20.5 Marketing process logical model.

approaches throughout this book to help make this part of the system easily understandable.

Process management systems architecture

Finally, of course, we must also deal with the issues involved in the design of the process management part of the system. While the rest of the marketing database world is best managed as a data warehousing kind of application, process management systems must, unfortunately, be developed just like any other on-line transaction processing (OLTP) system. This means that databases, platforms, and users are all subject to the whims and idiosyncrasies of whatever programming language has been selected for their development.

Hardware and software organization

Given this extremely complicated environment, then, it should come as no surprise that the actual architecting of a marketing database system in a telecommunications firm is a complex undertaking. Each part of the process—the acquisition, storage, access, and process management portions—must be supported by specific hardware, software, and processes that are uniquely defined according to the physical environment that the telco has created for itself. In general, then, those selections are made on a case-by-case basis.

Conclusion

With this chapter, we bring a close to our exploration of the many issues and concerns involved in the execution of effective, strategic marketing within the telecommunications firm. It is my hope that you have found many interesting and useful insights that will help you to build your own effective and efficient marketing database environment.

About the Author

Being both a consultant and a marketer at heart, I could not pass up the opportunity to tell you a little bit about myself, both so that you better understand the context from which this book is written (my consultant side), and so that you know the kind of work I do and how to find me if you need any help (my marketer side).

As a professional executive consultant for 15 years now, I have experience working with a variety of different telecommunications firms. In fact, my first "real job" in IT was as a full-time employee of AT&T in Chicago. Since then I have had the privilege of talking and working with people from telecommunications firms around the world, including Deutsche Telekom, France Telecom, Nippon Telephone and Telegraph, and many others. I have also had many engagements with U.S. and Canadian telcos including AT&T, Sprint, Bell Atlantic, Pacific Bell, SBC, United Telecom, Centel, Bell Mobility, and others. Currently, I work as a globally available executive consultant, trying to spread the good word about how best to build marketing database systems for telecommunications companies.

I most often begin by working with the computer systems personnel, but then end up spending considerable time with executives, especially from marketing, customer services, sales, advertising, and operations, trying to help them determine how best to coordinate the activities of large groups of people to more effectively perform the marketing function. In other situations, I am called on specifically to address issues about the best way to build and/or make use of marketing database systems that are already in place, but for some reason are not being utilized to their full potential. At times, I am asked to assist companies in their selection of hardware or software, design databases or systems, or to help make staffing and implementations decisions for many different kinds of telecommunications data warehouses.

In addition to these professional pursuits, I am also heavily involved in the activities of the National Conference of Database Marketing and in the International Engineering Consortium. I am always interested in opportunities to write about telecommunications marketing issues for magazines and other trade publications, as well as speaking at universities, conferences, and other gatherings of like-minded individuals.

I have written 10 books about databases, technology, data warehouses, Web technology, knowledge management, and telecommunications. Other Artech House publications include *Data Warehousing and Data Mining for Telecommunications* (1997).

Index

Teletraffic Technologies in ATM Networks, Hiroshi Saito

Understanding Modern Telecommunications and the Information Superhighway, John G. Nellist and Elliott M. Gilbert

Understanding Networking Technology: Concepts, Terms, and Trends, Second Edition, Mark Norris

Understanding Token Ring: Protocols and Standards, James T. Carlo, Robert D. Love, Michael S. Siegel, and Kenneth T. Wilson

Videoconferencing and Videotelephony: Technology and Standards, Second Edition, Richard Schaphorst

Visual Telephony, Edward A. Daly and Kathleen J. Hansell

Winning Telco Customers Using Marketing Databases, Rob Mattison

World-Class Telecommunications Service Development, Ellen P. Ward

For further information on these and other Artech House titles, including previously considered out-of-print books now available through our In-Print-Forever® (IPF®) program, contact:

Artech House	Artech House
685 Canton Street	46 Gillingham Street
Norwood, MA 02062	London SW1V 1AH UK
Phone: 781-769-9750	Phone: +44 (0)20 7596-8750
Fax: 781-769-6334	Fax: +44 (0)20 7630-0166
e-mail: artech@artechhouse.com	e-mail: artech-uk@artechhouse.com

Find us on the World Wide Web at:
www.artechhouse.com